# Too Much Is Not Enough

## *How Digital Technology Is Corrupting Society*

By

**Lt. Gen. Clarence E. McKnight, Jr.**

With

**Hank H. Cox**

ISBN: 1540643069
ISBN-13: 978-1540643063

# CONTENTS

Too Much Is Not Enough

# Chapter One

# Too Much Is Not Enough

*There seems to be an excess of everything
except parking space and religion.*
—Ken Hubbard

Mark Twain famously said that the trouble with whiskey is that too much is not enough. I believe the same phenomenon prevails in this modern information age, only I am not talking about booze (though God knows we have too much of that, too). Rather, I am talking about information – facts, data, intelligence, numbers, names, addresses, analyses, gossip, rumors, all the steady stream of information that comprises the information age rendered ever increasingly abundant by digital technology. We all are deluged with more information than we can process or use sensibly and effectively. There is no way to turn it off, not that anyone wants to. This is unfortunate because the surfeit of knowledge is rendering us ignorant and our nation less secure.

The tidal wave of information is borne aloft on the newfound power of digital technology that is rapidly infesting every aspect of modern life. There is a common presumption that the new digital technology presages a bright new era of progress and economic expansion such as ensued when electricity first came into widespread use, and without question it is having an impact. Every day brings reports of new developments made possible by technology – 3D printing, new pharmaceuticals to treat stubborn illnesses, advanced robots that perform a growing variety of tedious tasks to perfection, electric automobiles and trucks that drive themselves, machines that talk to each other in the "Internet of Things," ubiquitous security cameras to monitor systems and discourage crime, and remote control drones that perform all manner of complicated tasks in the workplace and on the battlefield.

The new digital age has dawned first in the good old United States which, for all its faults and shortcomings, is receptive to new ideas and generally optimistic about the possibilities of progress. Many cultures resist change and see innovation as a threat to eternal verities. In contrast, Americans are ready, eager and willing to embrace just about anything that comes in a pretty package and is touted in the media as "the next big thing." This is a positive asset in terms of its ability to enable progress, though it sometimes leads us down blind alleys.

The digital technology is everywhere and the information revolution is upon us, but our nation is mired in a bitter malaise born of despair that working people are losing ground and young people have meager prospects for the future. We have an advanced, well funded national security apparatus with its fingers monitoring human communication around the globe, but every fresh terrorist attack finds us flatfooted. Our young people have more information at their fingertips than we had access to long ago, but they somehow emerge from school uneducated, uninformed and ignorant of their heritage. The question is -- what good is this digital based information revolution doing us? Where is the productivity, the progress?

Despite this marvelous new technology, the U.S. economy continues to limp along. Our GDP growth is anemic by historical standards in large measure because of a lack of advances in productivity. People continue to die of cancer despite decades of research into cures, electric cars cost a fortune and will not go very far, robots too often run amok, and terrorists, domestic and foreign, continue their depredations despite our best efforts to identify them and deter their outrages.

The U.S. economy emerged from The Great Recession several years ago, but it has yet to manifest anything near the economic growth that characterized our economy throughout most of the latter half of the previous century. We seem to be trapped in an endless cycle of anemic growth in the 1 ½ to 2 ½ percent range. This is a bit better than what is going on in Japan and Europe, but not much. It is clearly not enough to validate the promise of prosperity for the vast majority of working people. The emerging digital technologies are producing a boatload of wet-behind-the-ears billionaires, but very little of that money is reaching down into the economy where it is vitally needed by people who are struggling.

The ramifications of this economic weakness are profound and raise serious questions about the future of our country. The so-called "American dream" is predicated upon the notion that everyone with gumption and a bit of luck could make it here. Year after year in the post World

War II period living standards went up. Young people coming out of school encountered myriad opportunities on every hand. Our world class factories offered wonderful opportunities to people with limited education and training. A factory job was a surefire ticket into the middle class. If you didn't make something of yourself in those halcyon days, you had no one to blame but yourself.

But the low skill factory jobs have gone away and we are left with a mostly service economy that seems built upon low wages and minimal benefits. Higher education was once a ticket to a better life, but our institutions are increasingly disconnected from the real world we live in. Today our country is flooded with college graduates tending bar and driving taxis, all too often saddled with huge debts they ran up to attend universities, debts that are not amenable to the bankruptcy code. The good jobs just aren't there but the debts remain.

It is a given among economists that economic growth depends on gains in productivity. But despite all this marvelous digital technology we are limping along at 1-2 percent growth rates which economists, led by Federal Reserve Chairman Janet Yellen, assure us is because of anemic productivity. There is a widespread assumption that the emerging digital technologies hold the key for renewed productivity growth, Economists ponder the vast new information revolution all around us and wonder why it is not boosting productivity.

There are conflicting opinions whether the new technologies actually presage a new era of growth. In his recent book, "The Rise and Fall of American Growth, The U.S. Standard of Living Since The Civil War," economist Robert J. Gordon contends the highly touted potential impact of the information revolution simply isn't there. He says the future is all too likely to be marked by stagnant living standards for most Americans because the effects of slowing technological progress will be reinforced by a set of "headwinds": rising inequality, a plateau in education levels, an aging population and other factors.

Gordon identifies five Great Inventions that powered economic growth from 1870 to 1970: electricity, urban sanitation, chemicals and pharmaceuticals and chemicals, the internal combustion engine and modern communication. He contends those things have had their impact and there is nothing on the horizon to replace them. I believe Gordon is too pessimistic by far because the real impact of the technological revolution is only now beginning to be realized. But his vision is sobering and should serve as a wakeup call. We need to recognize the challenge of making the promise of technology a reality.

For example, of the five great inventions Gordon cites, I believe electricity was the key driver. Electricity got its first commercial use in the telegraph in 1837, and by 1870 it was coming into use for street lights and industrial machinery, but it had yet to appear in private homes. It took about a century for electricity to reach its full potential in this country, and of course it has yet to reach many remote corners of the globe.

Why so long? Engineers had to resolve complex challenges posed by this revolutionary new energy source. Entrepreneurs fought for a piece of the emerging market. There was a long-running battle between alternating current (AC) and direct current (DC). (Thomas Edison was a diehard champion of DC.) Patent lawsuits proliferated. Government took forever to adapt with laws and regulations needed to assure safety, access and competition. Financial markets had to raise vast sums to fund generation plants and a network of distribution lines. Citizens had to get used to electricity and learn how to handle it.

Everyone with any sense could see that electricity would remake the world we live in, but it was not possible to imagine how it would all work out, anticipate the challenges and adjudicate the conflicts quickly. Also, it took a long time for industry to conceive and produce the myriad machines and consumer products that were made possible by electricity – and which boosted productivity and economic growth to unprecedented levels.

We are today wrestling with a similar challenge as we attempt to envision the potential applications of digital technology, anticipate the challenges and adjudicate conflicting interests in order to achieve its full potential. We are now caught up in the "gadget phase" of the digital revolution, awaiting the latest new tricks from Silicon Valley.

But gadgets are not the answer. In fact, I am beginning to believe they are more of a problem. People walking down the street or sitting in restaurants or riding in cars or watching movies in theaters are incessantly talking into invisible microphones or reading texts, borne along on an endless tidal wave of information oblivious to the world around them, unconcerned that their incessant chatter is mainly just noise annoying people around them. A growing number of these data junkies are being killed in car wrecks or run over in pedestrian crosswalks because they are riveted to their smart phones and unaware of the real world around them. Access to information is not always empowering people. Sometimes it is literally killing them.

There is a legitimate question whether all this data is

actually making us smarter, more productive and safer. Like booze hounds ordering another round and then another and another, we keep grasping for more and more information as if the very volume of information would make us wiser, richer and safer. Minute by precious minute our lives are consumed in a mad quest for ever more information because we realize instinctively that the information we have is not enough. So to resolve our quandary we keep grasping for more. It is accepted wisdom that the very definition of insanity is doing the same thing over and over again expecting a different result, but that is precisely the cycle we are caught up in.

The steadily declining test scores of school students who, through the magic of digital technology, have access to more information than their parents ever did suggests something is seriously amiss. For decades we have been assured by the experts that the Internet would open the door to a bright new age of universal access to knowledge and educational excellence, but the reverse seems to be the reality. Today's students have mounds of information at their disposal and ready access to any information they may desire, but they lack the wherewithal to handle it – to glean real world wisdom from the tide of information washing over them.

The technology becomes an end in itself, a digital Disneyland filled with fun rides to nowhere. The students squander billions of hours chatting and texting away, sharing rumors and gossip, to no coherent purpose. Much of the information they come across is bogus because there are no checks and balances on the Internet. When you read a newspaper, you can assume that some reasonably well educated people with professional integrity have processed that information and sought to verify its sources. But the Internet is populated with hucksters and hustlers who spread nonsense and rumors and conspiracy theories born of their disturbed imaginations that take on a life of their own. The younger generation feeds on this endless stream of mindless babble and accept it as reality. They believe it must be valid because they acquired it off the Internet through their digital devices that are always in hand, an ever present Linus blanket connecting them to an imaginary universe spinning farther and farther away from reality.

The young people are intensely busy with their digital gadgets but they are not achieving anything, not making progress, not getting ahead or moving up. It is a fair question whether they would be better off without those devices. The educators marvel at the potential of digital technology, but look on in despair at the decline of our public schools and wonder where it all went wrong. Many of the students appearing at four-year universities are woefully unprepared for college work but they still gain admittance because the schools desperately need those tuition dollars.

Our government offices and business enterprises are likewise caught up in the information revolution with personal computers on every desk and clever software controlling factory floor machinery, and yet somehow efficiency is languid and productivity growth is stagnant. Similarly, our national security data gathering apparatus is a sprawling octopus tapping phones and e-mails around the globe, assembling vast archives of information about terrorists bent on havoc, but the sheer volume of the information thus gathered mitigates against effective analysis and effective use of it. Every new incident of terrorism brings reports that there was solid information about the culprits in the system, but it was somehow overlooked. Of course it was overlooked. Almost all of it is overlooked because there are not enough qualified people to process it all. We had the information but it did us no good whatsoever. We are drowning in information, but the sheer volume of information does us more harm than good.

Perhaps most troublesome of all is the disturbing way digital technology is reconfiguring human relations as people become addicted to their digital media, abandoning the normal discourse of human interaction. I walk into cafes and coffee shops and see dozens of people around me intent on their cell phones, oblivious to the people around them. According to the research firm eMarketer, Americans spend an average of five and a half hours a day with digital media, more than half of that time on mobile devices. Among some groups the numbers range much higher. In one recent survey, female students at Baylor University reported using their cell phones an average of 10 hours a day. In a 2015 Pew survey, 70 percent of respondents said their cell phones made them feel freer, while nearly 30 percent said they felt like a leash. Nearly half of 18-29 year olds said they used their phones "to avoid others around you." We have unleased this powerful technology without thoughtful consideration of its implications.

The information revolution as it is now playing out is getting away from us, running amok like water from a burst dam, sweeping all in its path. We are overwhelmed by both the volume of information and the speed with which it moves. We are so busy gathering information we have little time to process it, winnow out the chaff and extract the knowledge hidden within. We are all busy wrestling with mounds of data and information when we should be sitting there thinking. I fear that to some extent young people are no longer learning how to think and we older folks are forgetting how.

In his seminal work "Amusing Ourselves To Death," first published in 1987, Neil Postman warned of a pending "information glut" well ahead of the digital revolution, and he was not optimistic about the potential consequences. "Our politics, religion, news, athletics, education and commerce have been transformed into congenial adjuncts of show business, largely without protest or even much popular notice," he wrote. "The result is that we are a people on the verge of amusing ourselves to death."

We need a recognition that the potential of the digital revolution is slipping away from us. The core challenge is using digital technology to boost productivity and the quality of human life – like electricity did. How do we quickly translate the creative genius of digital technology into useful applications in industry and commerce? How do we figure out how to use technology to help young people learn instead of distracting them? How do we manage the flood of intelligence data to mine the nuggets of real information? How do we harness the new power and bend it to our will? How do we employ the new technology to teach and reinforce fundamental democratic values? In sum, how can we grease the skids and accelerate the march of technology to enhance the quality of life and the security of our country?

We believe the technology revolution augurs great progress, but how can we step it up? This challenge is so big that government must take the lead – fostering broad-based collaboration of business, science, academia, entrepreneurs, engineers, government and risk capital to hasten the interconnection of all computer-based systems we depend upon, expanding the Internet of things, making fast Internet access available to everyone, and minimizing silos of proprietary software in favor of universal service and accessibility. In particular, we need to encourage more rapid infusion of advanced technology into our most stubbornly inefficient sectors – medicine, education and national defense – where it is urgently needed.

The first and most crucial task is to recognize the fundamental challenge – accelerating the process of integrating advanced technology into our economy to boost productivity. We must jumpstart productivity to get the economy growing robustly. Economic growth will pave the way to deal with all of society's ills – from income inequality to lack of opportunity to international competitiveness to national security. This will be a highly complex undertaking demanding buy-in from all sectors of society. In 2017, we will have a new President. He or she should immediately convene a high level commission representing all affected sectors to develop a comprehensive plan to accelerate the digital revolution, realizing its potential to boost economic growth. The electricity revolution took a century; we cannot afford to wait that long to realize the promise of the digital revolution.

# CHAPTER TWO

# The Bad Old Days

*When you have to make a choice,*
*and don't make it, that is a choice.*
—*William James*

When I was graduated from West Point in 1952, I had a range of options before me regarding what specialty to pursue in the military. Regrettably, one of them was not the Corps of Engineers which I had set my heart on many years before. In those days it was seen by most West Point graduates as the most promising route to achievement and higher rank. I am told that today more cadets are interesting in leading soldiers in battle, which I find heartening. As it happened, I graduated about mid pack of the class of '52 (206 out of 528) and there were 50 or more of my classmates ahead of me on the list for the Corps of Engineers.

Looking back today, I cannot recall if I chose the Signal Corps or the Signal Corps chose me, but it was to prove an excellent choice, whoever made it. I was of course fully aware of the importance of the Signal Corps. The most valuable asset on a battlefield is information – knowing where the enemy is, where our troops are, how the conflict is unfolding. Timely, accurate information is worth more than more troops and artillery. Then and now, in every military strategy and on every battlefield, information is critical to the outcome, usually the decisive factor.

In retrospect, my early career seems like a mad dash into the unknown. I got my sheepskin, married my wonderful wife Barbara and was soon on my way to war in Korea. (There was some additional training between West Point and the battlefield, but it did not deter me for long.) When I disembarked in Korea and was taken to my assigned unit as a lowly second lieutenant, I found myself running around under fire repairing the communications cables strewn between the central command and the front lines which were forever being severed by artillery rounds or North Korean troops slipping behind our lines with metal cutters. There were no satellite communications in those days. In fact, our backup plan for communication was based on carrier pigeons. On one occasion, I actually tested our pigeons sending a group of them with messages back to headquarters several miles away over those rugged, frozen Korean mountains. Most of them got through. Fortunately, we never actually used them in combat situations.

Over my 35 year career in the U.S. Army, I witnessed extraordinary progress in the realm of communications technology that was constantly challenging the Signal Corps to keep abreast of changes and make certain our military had the best possible communications. Using newly-developed vacuum tubes, we came up with small walkie-talkies that operated on FM, which was a tremendous advantage on the battlefield. The advent of transistors was another step forward as was the invention in 1958 of the integrated circuit, or electronic microchip. For years communication depended on wires, and we still have many wires transmitting telephone calls and cable programs around the world but now satellites spinning over our heads offer a whole new way of transmitting signals that do not require anything on the ground but receivers.

Along with revolutionary developments in communications technology came advances in computers. When I was posted to the Army missile range in White Sands, New Mexico, in 1961-62, I was put in charge of one of the first asynchronous computers, the Philco 2000, that took up an entire room. We thought it was a marvelous device and within the context of the times, it truly was. We used it mainly to project the trajectories of missiles, factoring in weather conditions. That was the last gasp of the age of transistors. We are now into digital technology. Today elementary school kids have more computing power in their backpacks than that Philco 2000. Indeed, we have more sophisticated technology in our cell phones.

We had access to marvelous breakthroughs in communications during my career in the military, but our all too frequent inability to take advantage of them should have served as a warning to what was to come in the greater society. By the mid-80s, I was wearing three stars on my collar as senior officer of the Signal Corps (we no longer had a commandant title – but it was me) and was posted to the Pentagon where like Sisyphus I spent a long time trying to roll that huge stone uphill persuading all of the branches of the military services to embrace the modern digital communications technology, and to learn a common language. It was a frustrating struggle. I love the military but it is by nature conservative, defensive of turf and resistant to new ideas.

But fate intervened. In the fall of 1983, two Shiite terrorists drove a bomb-laden truck to the barracks hosting U.S. Marines at the airport in Lebanon killing 241 military personnel in what proved to be the worst single day disaster for the Marines. In the aftermath, bitter inter-service rivalry paralyzed the Joint Chiefs of Staff which proved unable to come up with a retaliatory plan. While they bickered among themselves, injured Marines were flown back and forth all over Europe while the services argued about which would treat them. It was a miracle none of them died in transit. The lack of coordination among the services made the front pages and became a national scandal.

Two days later, U.S forces were sent into Grenada in the Caribbean to oust a Marxist government that was trying to create another Cuba. As happened all too often, the goal of the Pentagon was not to achieve a military objective, which was never in doubt, but rather to make sure all of the services got some of the action. So the Marines took the northern part of the island and the Army went into the southern part. Special operations were carried out by both the Army's Delta Force and the Navy's SEAL teams here and there. I don't believe the Air Force was actively engaged but it was there overhead in case a likely target presented itself.

Once again the lack of interoperability and cooperation reared its ugly head. None of the different services could communicate with each other. As senior communications officer for the Pentagon, I was more than embarrassed by the lack of interoperability of communications among the services. Army units were unable to communicate with Navy vessels to guide offshore supporting gunfire. Army officers even flew to the Navy command ship to coordinate their activities, and another Army officer used a Marine radio on the ship to call other Navy ships, but was unable to get through because he lacked the Navy codes to authenticate his requests. There was even one report of an Army officer using a pay phone on the island to communicate with his superiors back in the states since he could not communicate with anyone at the scene. Since each service bought its own communications equipment independent of each other, there was little interoperability.

In sum, it was a classic Chinese fire drill of miscommunication and sometimes total lack of communication that would have spelled disaster had we been up against a more formidable foe. It was clear to me and many others that we had to find a dramatic solution to this mishmash of competing systems. All of which became part of an ongoing debate about confusion and miscommunication among military branches that set the stage for the Goldwater-Nichols Act which forced the services to start talking to each other and working with each other. I worked closely with the Senate on that critical legislation, running back and forth from Capitol Hill to the Pentagon until we got a bill we could work with. I can testify that there are easier tasks than trying to reconcile opposing views of Senators and military brass, and there were times I thought my sterling career was going down the tubes, but I believed what we were doing was vital to the security of our country so I stuck with it. The final bill led to a lot of grumbling in the Pentagon, but it achieved our primary goals and was a positive gain for the nation.

In those years, as part of my campaign for interoperability and modernization, I spent countless hours working to nurse the Army and other services along from analog into the digital age that took communications to a higher level, providing more efficiency and security. But it was a tough slog. The basic problem was a generational divide that has hampered our society for years. The older generation of leaders in both the private and public sectors simply does not understand the new technology and is not comfortable with it, but they are the people in charge who must make the decisions. There were times I had to send young wet behind the ear second lieutenants to lecture senior officers about computer software. That doesn't go down well with the brass, and it did not make me popular among my peers, but the young people understood what we were dealing with and were not wedded to the old ways.

Digitization quite simply makes it much simpler to hook things together over longer distances. I retired from the Army in 1987. It was my regularly scheduled end of duty but I was not happy about it knowing the job I was doing was far from finished. Of course, as always there were other capable people to step up and continue the work. I proudly saw the results of my handiwork – and that of many others, of course – during the 1991 Gulf War (Operation Desert Storm) when the nightly network TV news showed clip after clip of officers huddled over laptop computers before a giant video screen, closely managing the complex movement of a massive military operation covering thousands of square miles of land and sea. That was real time communications in action. I do not believe any military had ever gone to war with a more closely synchronized operation that was made possible by an incredible communications network.

## Into The Private Sector

Upon retirement, I took a position with Booz Allen Hamilton, the prestigious consulting firm that did extensive support work for both the military and the commercial sectors. By 1992, I was working closely with the Defense Advanced Research Projects Agency (DARPA) that had been created in 1958 to help make sure the U.S. maintained its technological leadership over the Soviet Union during the Cold War. DARPA employs some of the brightest people in the nation working on projects that have great impact on our future. It was DARPA that invented the Internet, initially as a means for academic researchers to share information. Of course, those super bright people tend not to stay with DARPA very long because the private sector beckons with more lucrative opportunities, but they are invariably replaced with even brighter people. I got to know many people at DARPA and to appreciate their work when as an active officer I was commanding Fort Gordon in Georgia and Fort Huachuca in New Mexico, mainly working with them on innovative concepts for packet and digital radios.

I should interject here that I have had a lifelong interest in educational programs for our nation's youth, particularly in math and science. It was education that enabled me to rise from humble beginnings and over my long career I began to perceive a diminishing education level in recruits to the Signal Corps. We had to do a lot of training to compensate for inadequate preparation in the public schools.

There was in those days no concept of STEM (science, technology, engineering and mathematics) as a distinct area of study. That would come later. One of my first initiatives in this area was one that I inherited from a former commander of Fort Gordon when I was a lowly colonel, Major General Bill Hilsman. He envisioned an educational center for youngsters to expose more of them to exciting things going on in modern digital technology. This was a time when the great majority of public schools, even more hidebound than the military, were oblivious to the digital revolution always well underway. After a few fits and starts, it morphed into the National Science Center. Soon we had two big 18-wheel trucks, called our Mobile Discovery Centers, hauling our science and technology around to public schools near and far. The purpose was to teach young people that studying science and math is fun, as well as essential to their future. Each unit included a mobile theater seating 30-35 and showing video programs about math, science and computers; a mobile classroom providing training for students and teachers; and a mobile walk-through exhibit showcasing small interactive exhibits and displays.

We wanted to build a permanent National Science Center there at Fort Gordon in Augusta but could never round up the funding. But through this outreach to young people with the trucks, I became convinced that they were stuck in the same old public schools doing the same old routines at the beck and call of teachers who were also trapped in a time warp, oblivious to the rapidly changing world around them. I believed we were missing a great opportunity to use the new technologies to improve the public schools – not least by bringing the new technology into the schools and training teachers to work with it.

At the time, I was working with DARPA on some

project, I forget just what, and my contact there was a bright young fellow named Mike Kelly who shares my background in electrical engineering. Mike had a sterling career at IBM and has taught at several top rank universities. At the time, Mike was mounting a crusade to get a computer on the desk of every K-12 student in the country, an ambitious proposal that many thought both desirable and doable.

This was in the early 1990s. In those days, Japan was running circles around us in technology and our government was increasingly concerned about our loss of leadership in technology. In 1987, Congress created SEMATECH (Semiconductor Manufacturing TECHnology), a government-sponsored consortium of 14 leading U.S. semiconductor manufacturers to enable them to pool resources and knowledge in order to compete more effectively against the Japanese. DARPA was the conduit through which Congress provided $500 million to SEMATECH. Kelly was involved in that operation born of his lifelong concern about the quality of education in our country – an interest I shared. The semiconductors manufacturers, of course, were motivated by the prospects of profits – as befits private sector companies. But the concept of a computer on every school desk was part of the SEMATECH vision.

The plot thickened. Mike put me in contact with a real life tech entrepreneur named Jack Taub, now deceased, who had helped launch one of the first online networks, The Source, which he later sold for a substantial amount of money. (The Source morphed into CompuServe after which the evolution gets cloudy but of course we now have many competing online platforms.) Jack shared our inspiration that digital technologies had a great potential to transform K-12 schools. At the time, Jack was working with a young vice president of the U.S. Chamber of Commerce named Jeffrey Joseph who also had a vision of the power of technology to transform education, reflecting growing anxiety among business leaders that our young people would not be able to function productively in the workplaces of the future.

But I hasten to stress that this was never a vision that embraced technology as a cure-all for our educational challenge. "I have been consistent in trying to get technology into the classroom," Kelly said. "But the most important element is the teacher. Today, one third of all teachers come from the lower third of their graduating class. This has to change. I compared this with what they were doing in Finland where only graduates in the top tenth are accepted into teaching. The most important aspect of improving education needs to be preparing teachers to use the technology."

"There is a growing body of evidence from research, schools and workplaces in a variety of settings around the country that sophisticated communications technologies offer the promise of quantum advances in education and learning through more individualized instruction," Joseph told the House Committee on Economic and Educational Opportunities in 1995. "In part this is simply because modern students and workers are more acclimated to using advanced technology than earlier generations were, and commensurately less receptive to the old, pedagogical teaching technique."

Joseph spelled out the promising of using digital technology to empower teachers. "Properly used, the high tech classroom can accommodate all students at their individual speeds and needs," he said. "Thus, teachers will no longer be required to seek out a middle ground of progress, leaving slow students behind and boring brighter ones. By merging interactive technologies with other teaching techniques, instructors can accommodate each students' abilities and interests."

Right there in the library of the U.S. Chamber of Commerce on H Street in Washington, D.C., directly across Lafayette Park from the White House, Joseph had assembled a demonstration module of the most advanced digital communication technology that included online data access and interactive videos, all the bells and whistles of the modern digital age. I could see in an instant that this represented the culmination of the evolution of communications technology up to that time, at least everything I was aware of, everything I had been working toward in my career up to that point. The question before me was how do we get from here to there? How do we translate this vision into reality and make it available to everyone?

# CHAPTER THREE

# Launching Digital Learning

*Computers are useless. They can only give you answers.*
—*Pablo Picasso*

My connection with Jeff Joseph was born of our shared interest in educational reform. The U.S. Chamber of Commerce where he served as vice president for many years had surveyed its members in the late 1980s and discovered a wellspring of concern about education. Even way back then the young people coming into the workforce simply were not as well prepared as earlier generations had been. The public schools were getting out of the vocational arts business, and generally not replacing it with anything better. For the most part, they were oblivious to the digital revolution just then getting legs. The business community was growing concerned that over time this issue, left unaddressed, could become a major impediment to the nation's competitiveness.

All of this, of course, was an echo of "A Nation At Risk," the 1983 commission report fostered by the Reagan Administration that branded the poor quality of the nation's schools as a threat to the future. "The educational foundations of our society are presently being eroded by a rising tide of mediocrity that threatens our very future as a Nation and a people," the report concluded, and "If an unfriendly foreign power had attempted to impose on America the mediocre educational performance that exists today, we might well have viewed it as an act of war." That report has led to many dramatic efforts to improve the quality of our schools, and expenditure of vast sums of public and private money, thus far to little avail.

But back then it looked like a tractable problem just waiting for some serious investment and clever leadership. The business community was an obvious place for a reform movement to get started. Thus, Jeff had launched the Center for Workforce Preparation and Quality Education within the Chamber. Jeff was relatively young by my standards, still in his early 40s, but he had made a name for himself at the U.S. Chamber as the point man on a number of critical issues. He had debated Ralph Nader in a National Town Hall meeting hosted by National Public Radio, was a frequent witness before Congress on a host of controversial issues, and had many connections with the Bush Administration. He had a great rapport with U.S. Chamber senior management and the board of directors, and thus was able to create and staff the Center. On his own initiative, he had resolved to make educational reform a U.S. Chamber priority and he had the clout to do it.

Jeff went about shaking his many trees of connections looking for people who might be able to assist in his education reform campaign. A mutual friend suggested – insisted really – that Jeff meet with Jack Taub. Mentioned earlier, Jeff had never heard of Taub but to cognoscenti of the digital evolution, Taub is a familiar name. He was one of the first to realize the potential of the new digital technology. He launched the first online service, The Source,that he later sold for a substantial sum. The Source morphed into CompuServe which eventually was absorbed by one or more of the developing online services. With his fortune made, Taub was casting in those days about for creative ways to employ the new digital technology in the public schools to capture the imagination of young people and revive the quality of education at the K-12 level.

After checking around about Taub and confirming Taub's bona fides, Jeff brought him in to discuss education. Soon they were off and running, two visionaries with big ideas for changing the world and little hesitancy about trying. Taub introduced Jeff to many people involved in the early stages of the computer revolution, among whom were Mike Kelly and another guy named Jim Carey who also worked for DARPA.

Carey and Kelly told Joseph that DARPA had huge sums of money to invest in technology mainly at the behest of then Senator Al Gore, a Democrat from Tennessee, and Senator Jeff Bingaman, a Democrat from Arizona. Somewhere along the line Jeff decided a military connection would help with his vision of a major campaign for educational reform which led to Carey and Kelly introducing him to me. Soon all of us – Carey, Kelly, Taub, Joseph and me plus a few others were discussing visionary schemes to get computers in every school and on every desk. We were free lancing really, acting independently, not part and parcel of any organization or government agency. That gave us a certain amount of freedom, but also deprived us of organizational support.

But we had resources. With astounding quickness, Joseph was receiving money from something called Delta Research for the Chamber's Center and soon all of that technology was being set up in the U.S. Chamber library on H Street, across Lafayette Park from the White house. It was all there. It would seem commonplace today but in 1992 it was revolutionary stuff. You could conduct face to face meetings with people in remote locations and access remote data bases. Who ever heard of such things? The future was right there before our eyes. It was impossible not to get carried away.

The actual source of the funding was somewhat cloudy. "It wasn't actually DARPA money, but rather money from one of those off budget accounts that bedeck the Defense Department," Joseph said. "I was told it was money left over from the campaign to oust the dictator Noriega from Panama. I could never prove that of course. It was orphan money, but it was real enough."

Unfortunately, the fact that any government money was coming into the building was disturbing to the senior brass of the U.S. Chamber of Commerce which was and remains highly suspicious of the government. Joseph was being asked many questions about it that he was reluctant or unable to answer. Though Jeff had risen high in an organization populated by ideological conservatives, he himself has always been a registered independent and aloof from any ideological persuasion. He is like me in that he has always been focused on what works. He was fascinated by what he had there in the Chamber library, paid for by mysterious government money, and it was in fact a wonder to behold.

"We had disparate systems there," Joseph said. "We

had IBM, Apple, Hughes, everything that was around back then was included. We needed a digital satellite. The Chamber had one, but it was analog. Hughes put in a digital dish and showed us how to make that system talk to other systems."

The Chamber was no stranger to technology. Under its then President Richard L. Lesher, it had invested heavily in television. The U.S. Chamber had its own TV studio in its headquarters building where it produced business news programs carried nationwide on ESPN. This was the forerunner to CNBC, Fox Business, MSNBC, all the business networks of today. Joseph was a regular commentator on *Nation's Business Today*, the Chamber's five-day a week business report carried on ESPN in the mornings.

Joseph had big ideas for what to do with the technology. He and Taub aspired to wire up all of the nation's schools and also provide online learning to adults. Joseph wanted to create a one-stop shop for businesses to acquire information online, cementing the U.S. Chamber's role as leader of the business community. It was still the era of the first Bush Administration when George Herbert Walker Bush was the chief executive. Joseph was close with the Bush Administration as was the U.S. Chamber. Joseph invited senior people from the Bush Administration over to see his high tech demonstration model. But they were appalled by the implications for education. The President's advisors told Joseph it ran counter to their conviction that quality education was dependent on the church and family and could not be usurped by technology. "They just didn't get it," Joseph said ruefully.

Meanwhile, DARPA had its fingers in a variety of high tech enterprises and was sending a steady stream of visitors to the U.S. Chamber library to observe the demonstration module on display. At the time, Joseph greeted them as nice people who came to see his new toy and he had no way of knowing some of them would figure prominently in the digital revolution to come. Larry Ellison, co-founder of Oracle, was visibly impressed. "He basically told me that he was going to steal our technology," Joseph said. "He made no bones about it. There wasn't a lot I could do about it because I did not own the technology."

Another interesting visitor was Melissa French from Microsoft which was specifically interested in software for business. French would a few years later marry Bill Gates of Microsoft and the rest is history. "She was going to get back to me," Joseph said, "but we never heard from her again. However, we did see evidence of our influence in a lot of what Microsoft came out with in ensuing years."

This was a fascinating time in my life as we were frolicking on the leading edge of the digital revolution meeting people who were making history. We even met a guy named Doug Blankenship who had invented the computer mouse. As he explained it to us, the new technology was meeting a wall because senior executives and military officers refused to use computers because it required them to type. In their culture, serious leaders did not type. They delegated that to secretaries. But they could point without losing face. The mouse was invented so great mucky-mucks could use computers by pointing instead of typing.

All along my primary interest in what Joseph was doing and his demonstration module was its potential to enhance education. In those days, the new technology was only beginning to make its way into daily life – business, medicine, entertainment – and education was lagging. It only made sense that it would because schools are conservative organizations operating with tight budgets. Few if any people in education had a clue how the new technology could be used to enhance the learning experience. There was no budget for it and no one in the school systems who understood it. There would have to be an outside force bringing it into the classroom.

But DARPA understood the potential implications of the new technology and was willing to put up seed money for our vision. The result was the Community Learning and Information Network (CLIN) which was envisioned as a catalyst to enable community groups to come together within the context of private sector networks. Our original vision was that the private sector would provide the basic information tools and technologies for the new system and once that plan was operational, CLIN would be dissolved.

There were some business leaders more receptive to the concept than educators, especially at Booz Allen where I was then employed. In 1992, the U.S. Chamber of Commerce and DARPA launched CLIN as a joint initiative to enable a "culture of competence" across the nation creating an infrastructure that would be available. We were soon operating out of office on K Street.

By then, the Clinton Administration was in office and was clearly more technology friendly than the Bush team. They were amazed, in fact, to discover the White House was still using rotary telephones and had switchboard operators putting through calls like in the 1950s.

Vice President Gore who had come by the U.S. Chamber library to see the module and spoken at length with Joseph about its potential was definitely on our team. We asked DARPA for $90 million to get CLIN fully mobilized and were told we might actually get $30 million, which still seemed like a windfall to us.

Then politics got in the way, as it often does in Washington. As a separate issue, the U.S. Chamber of Commerce had been in negotiations with the White House about health care reform, but when Congress took it up, the U.S. Chamber reversed course and went all out to defeat it. The U.S. Chamber's legislative efforts, as always, were led by the same Jeff Joseph who was working with CLIN. As soon as the White House identified this connection, the DARPA money disappeared.

## The National Guard

Undaunted, Joseph bypassed the White House and went to Capitol Hill in search of political support. He found it with Senator Robert Byrd (D-WVA), the wily Mountain State legislator who was ever and always on the lookout for creative ways to channel federal money to his constituents. Senator Byrd inserted a $7.5 million earmark for a regional distance learning project that would see a pilot program in West Virginia.

The focal point of this distance learning project was the National Guard. The National Guard, the oldest military institution in the U.S., was challenged by its responsibility to maintain a combat ready military and emergency force with diminishing resources. Through Senator Byrd's office, we offered the CLIN vision as a practical solution to its dilemma. Using the technology we had developed, the Guard – and later the U.S. Army Reserves as well – could provide onsite interactive training in local communities where its members lived, saving a lot of money by not having to move people around the country.

No sooner did we get the language in appropriations legislation than we had a visit from a high ranking officer from the National Guard who was adamantly opposed to the project. He said the Guard would refuse to work with us and would tell Congress to keep the money. I was never sure whether this intervention was ordained by the White House or if it was just another case of the old Washington "not invented here" syndrome. In any event, I made a few calls to the Pentagon and Jeff made a few calls to Capitol Hill, and we eventually persuaded the Guard to take the money and launch a distance learning system.

Of course, the Pentagon had to put its own label on the program which became the Distributive Training Technology Project (DTTP). The original $7.5 million earmark funded demonstration sites in West Virginia, Virginia, Maryland and Pennsylvania. It proved popular and successful and grew by leaps and bounds. By 2002, we had $300 million invested in 310 training sites in all 50 states plus four territories.

These high tech communication centers have proven their value time and again. For example in the wake of 9/11, as rescuers rushed to assess the devastation in Manhattan, New York Army National Guard officials used the DTTP classroom capabilities to create a point-to-point audiovisual link with state emergency officials to help assess the status of gas lines, electricity and telephone connections. Guard command staffs in New York, New Jersey and Connecticut used the DTTP's continuous video teleconference capabilities to conduct briefings and coordinate their rescue efforts for everything from troop deployments to providing meals.

Today, the DTTP is the world's largest distance learning network enabling the Army National Guard to meet its education requirements by linking 3,200 armories and supporting multimedia computer training for 362,000 citizen soldiers. The Army National Guard has deployed more than 200 distance learning classrooms nationwide. In addition to armories, the classrooms are located in local schools, libraries and community centers to ensure that access to these resources is available to the broader civilian population. The centers are run on a fee-for-service basis so that states can rent network access to federal agencies and the private sector on the 28 days a month when the Guard is not using them.

This system was to my knowledge the very first online interactive learning system of any type and we had visions of greatness. We (CLIN) aspired to use these National Guard/Reserve centers as prototypes for the nation's public schools, showing the world how online learning could transform the educational process. It was and is an exciting vision with great promise.[1]

But we ran into some unforeseen hurdles that conspired against our vision, or at least have delayed its realization. The first and most obvious is a wall of indifference in the educational establishment that is actively hostile to outside input, especially from sources not traditionally seen as part of that establishment. There was also widespread ignorance of and resistance to digital technologies that is

---

[1] In developing this online technology, CLIN won approval of several patents that in our view embrace just about everything being done today in terms of online education and training. The matter is currently before the courts.

still conspicuous among educators.[2]

The second hurdle was a rush of private for-profit companies using online technologies to market education and training, many of which we believe were and are using proprietary technology that we developed. Widespread abuse of this system, much of it funded by government education loans, has sparked public concern and review by Congress. Our vision had always been for using technology to promote learning on a not-for-profit basis, but the for-profit sector had more resources and beat us to the punch, at least in terms of some important target audiences.

Last but not least was a decision by former Defense Secretary Donald Rumsfeld to finance the war against Iraq on the cheap by pillaging the National Guard's resources, including the funding needed for the DTTP. This is a topic that is much bigger than its impact on the DTTP, and beyond the purview of this book. But it is an indicator of the kind of snags and hurdles that have sabotaged the potential of digital technology to address our challenges.

I still see a great potential for a National Science Center in Washington, D.C., and the use of CLIN technology to enhance the quality of K-12 education. I hope I live long enough to see these things happen. But the overall picture of the digital revolution to date is that of a revolution run off the rails, obsessed with trivia and useless information, unable to focus on what is needed.

---

[2] The New York Times of January 4, 2012, carried a front page story about teachers in Idaho refusing to adapt to computer technology as required by a new state law.

# Chapter Four

# The Big Data Phenomenon

*Water, water everywhere, nor any drop to drink.*
*—Samuel Taylor Coleridge*
*The Rime of the Ancient Mariner*

I had thought the first time I encountered Jeff Joseph's DARPA-funded digital laboratory set up there in the U.S. Chamber of Commerce library on H Street across Lafayette Park from the White House that I had seen the future. In retrospect, I am inclined to think it was more like Dr. Frankenstein's laboratory. Jeff, Taub, DARPA, I and the other architects of that visionary system, like the legendary Dr. Frankenstein of fiction, had the best of intentions. We wanted to usher our nation into a new era, to empower the American people with a dramatic new technology that would remake our lives in wondrous ways. What we got was a runaway train going 100 directions at once, only a few of them worthwhile.

For me the most obvious shortcoming of the new technology was manifest in national security where I have spent my career. The capabilities of digital technology should have set the stage for leapfrog advances in national security. It empowered us in marvelous ways to seek out potential threats to our country and to better prepare our national defense. But the defense-national security establishment, aided and abetted by our representatives in Congress, mightily screwed it up, and I am left to wonder if we are not less secure than we were before the computer age dawned.

The trauma of 9/11 had far-reaching impact on our national security apparatus as Congress and the White House scrambled to reassure the American people that our government would respond to the threat and prevent such horrors from occurring again. I must concede that thus far there have been no further acts of terrorism of similar magnitude, which suggests we must be doing something right. But that doesn't mean the nation is secure, or as secure as it should be. Indeed, there are weak places in our defense that are vulnerable to future acts of terrorism.

One of the most obvious areas of concern is the Department of Homeland Security (DHS) itself created as a direct result of 9/11 in an effort to streamline and coordinate the array of federal agencies responsible for various aspects of national defense. One would have to spend many years in the nation's capital, as I have done, to appreciate the chaos that resulted from that seemingly sensible initiative. Bureaucracies by their nature are inflexible and resistant to leadership and change. To forge dozens of unrelated agencies overnight into one coherent entity would have been beyond the powers of the most visionary leadership.

And our leadership was anything but visionary. The driving thought then and now was simply to make a big media splash, create something really big and spend tons of fresh money. That is a concept Congress can handle and the bureaucracy can feed on. The result is a Rude Goldberg monstrosity of a department that, as Churchill said of a poorly prepared pudding, has no theme. Predictably the senior bureaucrats responsible for this great paper shuffle used it as an opportunity to get rid of poor performers from their own agencies of which there are many in the bureaucracy. This is a time-honored custom. As a general rule it takes an act of God to get rid of incompetent, nonperforming bureaucrats. Reorganizations such as that creating the DHS present a once in a lifetime opportunity for senior managers to clean house. The result was a lot of bureaucratic deadwood piled up at DHS.

As if that weren't challenging enough, no fewer than 22 committees of Congress have some level of jurisdiction over DHS, most of it stemming from before the agencies were combined into DHS. Obviously, just as Congress acted to unify the various agencies it should have also unified Congressional oversight into two committees – one in the Senate and one in the House – but that did not happen. That would remove opportunities for members of Congress to seize the spotlight and issue press releases about their fine work defending the country. So the plethora of committees continue to perform their antic rituals.

The overall result is a display of conspicuous inefficiency as many senior DHS officials are required to spend a good bit of their time responding to Congressional inquiries or testifying before committees. As one who has been through it, I can testify that preparing testimony is time-consuming and rarely if ever productive. You work through the night to prepare your remarks and anticipate questions only to see panel members wander in and out chatting with reporters or staff, oblivious to what you have to say. The only way to make the evening news is to blurt out something truly stupid and most of us veteran Washington hands are ever and always alert to that danger.

When I helped the Pentagon work its way through the Goldwater-Nichols 1986 defense reform, we strove to keep the system as simple as possible. The Joint Chiefs forged a national defense strategy based on nuclear deterrence, control of the seas and freedom of outer space. We did not at that time anticipate the arrival of the digital revolution that is remaking every aspect of the way we live and work, and posing unprecedented challenges to national security. But the work we did should have made it easier for the Department of Defense to embrace advancing technology.

We live in a digital world and national defense depends on our ability to master digital technology, anticipating and fending off cyber-attacks and overt terrorism. A major concern, one of many actually, is the ability of terrorists to obtain computer access to our "consequential infrastructure," the IT systems of private companies and utilities which, if manipulated, could cause a catastrophic event harming masses of people and wreaking economic chaos. One recurring nightmare for example is the prospect of terrorist hackers sabotaging our electrical grid. To bring down our electrical system for weeks or months would have cataclysmic consequences for our country.

We need a coherent national strategy to synthesize a digital element into our defense plans. What we have instead is bumbling bureaucracy, overlapping jurisdictions and, in general, chaos.

## Big Data

As one who has spent a goodly portion of his adult life raving about the folly of amassing much more information than we can analyze and use effectively, it will come as little surprise to learn that I am skeptical about the so-called "big data" phenomenon. It is the "information glut" Neil Postman referred to in "Amusing Ourselves To Death," times ten to the tenth power.

Big data is an all-encompassing term used to describe data sets so large and complex that traditional data processing techniques cannot handle them. To be sure, digital technology has endowed us with a massive capacity to accumulate data. Since the 1980s, our capacity to store information has roughly doubled every 40 months or so. As of 2012, every day some 2.5 new exabytes enter the system. An exabyte is one quintillion bytes. My eyes tend to glaze over when I encounter numbers like that. It's like reading that some constellation is 90 million light years from the Earth or that the dinosaurs roamed 160 million years ago. Such numbers are numbing.

Suffice it to say that an exabyte is a large number and when you assemble lots and lots of them, you have big data. We are talking about data sets so large and complex that traditional data processing applications are inadequate. There is a widespread and in my opinion pathetically optimistic assumption that the sheer volume of big data will lead to better decision making resulting in greater efficiency in a broad range of applications in business, science, government, public politics, service organizations, national defense, everything.

You can count me among the skeptics. Big data is by definition raw data – information gleaned by computers without rhyme or reason, verification or interpretation. Computer scientists, physicists, economists, mathematicians, political scientists, bio-informaticists (whatever they are), sociologists, and every other kind of "ist" are demanding access to the massive amounts of information being produced about people, things and their interactions. Everyone wants in on the action of the "next big thing."

But raw data is just what it sounds like – raw. It is just numbers reflecting actions and transactions without context or reasoned analysis. Precious few people know how to glean useful knowledge from the mass of information we are accumulating. No matter how comprehensive or well analyzed, said the *Harvard Business Review*, big data must be complemented by "big judgment" or it will do more harm than good. But big judgment is hard to find.

Data sets are growing exponentially because they are being gathered by cheap information-sensing mobile devices, aerial (remote) sensing, software logs cameras, microphones, radio-frequency identification (RFID) and wireless sensor networks. The world's technological per-capita capacity to store information has roughly doubled every 40 months since the 1980s. As of 2012, every day 2.5 exabytes of data are created. All of which creates a tremendous potential for abuse, misinterpretation and assorted mischief. Figures don't lie but liars will figure and in this blinding maze there is tremendous potential for mischief.

One of the biggest accumulators of big data is the National Security Agency (NSA) based in Fort Meade, Maryland, south of Baltimore. Now part of the seemingly endless DHS, NSA intercepts telephone and internet communications of over a billion people worldwide, seeking information on terrorism as well as foreign politics, economics and "commercial secrets." It is an open secret that NSA is running amok gathering data often in violation of legal restrictions. In a declassified document it was revealed that 17,835 phone lines were on an improperly permitted "alert list" from 2006 to 2009 in breach of legal restrictions, which tagged these phone lines for daily monitoring. Eleven percent of these monitored phone lines met the agency's legal standard for "reasonably articulable suspicion" (RAS).

A dedicated unit of the NSA locates targets for the CIA for extrajudicial assassination in the Middle East. The NSA has also spied extensively on the European Union, the United Nations and numerous governments including allies and trading partners in Europe, South America and Asia. This is all pretty much standard national security activity.

But NSA goes overboard. It tracks the locations of hundreds of millions of cellphones per day, allowing them to map people's movements and relationships in detail. It reportedly has access to all communications made via Google, Microsoft, Facebook, Yahoo, YouTube, AOL, Skype, Apple and Paltalk, and collects hundreds of millions of contact lists from personal email and instant messaging accounts each year. It has also managed to weaken much of the encryption used on the Internet (by collaborating with, coercing or otherwise infiltrating numerous technology companies), so that the majority of Internet privacy is now vulnerable to the NSA and other attackers.

Domestically, the NSA collects and stores metadata records of phone calls, including over 120 million

U.S. Verizon subscribers as well as Internet communications, relying on a secret interpretation of the Patriot Act whereby the entirety of U.S. communications may be considered "relevant" to a terrorism investigation if it is expected that even a tiny minority may relate to terrorism. The NSA supplies foreign intercepts to the Drug Enforcement Administration (DEA), the Internal Revenue Service (IRS) and other law enforcement agencies, who use these to initiate criminal investigations. Federal agents are then instructed to "recreate" the investigative trail via "parallel construction." It's open season at NSA. They reach out and grab anything and everything.

NSA also spies on influential Muslims to obtain information that could be used to discredit them, such as their use of pornography. Most of the targets, both domestic and abroad, are not suspected of any crime but hold religious or political views deemed "radical" by the NSA. According to a report in *The Washington Post* in July 2014, 90 percent of those placed under surveillance in the U.S. are ordinary Americans, and are not the intended targets. The newspaper said it had examined documents including emails, text messages, and online accounts that support the claim.

Many critics in the media and academia have been raising concern about NSA's extensive reach and access to personal information. In 2013, the extent of NSA's secret surveillance programs was revealed to the public by a disgruntled employee named Eric Snowden. According to Snowden's leaked documents, the NSA intercepts the communications of over a billion people worldwide, many of whom are American citizens, and tracks the movement of hundreds of millions of people using cellphones. Internationally, research has pointed to the NSA's ability to surveil the domestic internet traffic of foreign countries through "boomerang routing".

In my opinion, Snowden is a traitor who has compromised our nation's security and put many dedicated agents at risk. My concern is not about NSA's potential violation of privacy – we are at war and security agencies have a responsibility to ferret out terrorist conspiracies. But rather I believe NSA is so committed to gathering data here, there and everywhere it is drowning in it, retrieving more data than it can handle or make effective use of. I served in the Signal Corps for many years, and was its senior officer near the end of my career. I learned the hard way that gathering raw data is a poor substitute for gathering intelligence. From mounds and mounds of data, a few clever people can through due diligence glean a few actionable facts, and from those scarce facts a further, more intense gleaning may produce a few drops of genuine wisdom. But to amass vast amounts of raw data is not only a waste of resources, it clogs up the system and conveys a false sense of security.

Within the context of Big Data we have a new concept, something called "data mining," which sounds like just what the doctor ordered, but thus far it has produced mostly limestone and precious few golden nuggets of actionable intelligence. The NSA and FBI have so much Big Data backed up in their warehouse that their miners have little success in finding the gold. As a result, the agencies usually show up "a dollar short and a day late" when terrorists do their ugly work.

Researchers Danah Boyd and Kate Crawford say big data is less about data than about a capacity to search, aggregate and cross-reference large data sets. "Like other socio-technical phenomena, Big Data triggers both utopian and dystopian rhetoric," they said. "On one hand, Big Data is seen as a powerful tool to address various social ills, offering the potential of new insights into areas as diverse as cancer research, terrorism and climate change. On the other, Big Data is seen as a troubling manifestation of Big Brother, enabling invasions of privacy, decreased civil freedoms and increased state and corporate control."

I would go further. All information taken out of context is invariably misleading. By its very nature, the raw content of all big data is taken out of context. This is not a positive use of digital technology. It is technology running amok.

# CHAPTER FIVE

# Where Is The Economic Impact?

*Life fundamentally improved between*
*1870 and 1940 in a way that it hasn't since.*
—*Paul Krugman*
*New York Times columnist*

Electricity got its first commercial use in the telegraph in 1837, and by 1870 it was coming into use for street lights and industrial machinery, but it had yet to appear in private homes. It took about a century for electricity to reach its full potential.

Why so long? Engineers had to resolve complex challenges posed by this revolutionary new energy source. Entrepreneurs fought for a piece of the emerging market. There was a long-running battle between alternating current (AC) and direct current (DC). (The celebrated inventor Thomas Edison was a diehard champion of DC.) Patent lawsuits proliferated, consuming vast amounts of money and time that could otherwise have been invested in developing the new technology. Government took forever to adapt with laws and regulations needed to assure safety, access and competition. Financial markets had to raise vast sums to fund generation plants and a network of distribution lines. We needed FDR's New Deal with its rural electrification to extend the power lines to the farms and villages where people still read by oil lamps and hauled water from the well by hand.

Everyone with any sense could see that electricity would remake the world we live in, but it was not possible to imagine how it would all work out, anticipate the challenges and adjudicate the conflicts quickly. Also, it took a long time for industry to conceive and produce the myriad machines and consumer products that were made possible by electricity – and which boosted productivity and economic growth to unprecedented levels.

It was that boost of productivity that underscored the rising prosperity of our nation. It is an established fact of economics – to the extent that there such things as established facts of economics – that gains in productivity are the key ingredient of economic expansion. "The most important factor determining living standards is productivity growth," said Federal Reserve Chairman Janet Yellen in 2015, "defined as an increase in how much can be produced in an hour of work. Over time, sustained increases in productivity are necessary to support rising incomes."

And we have a serious productivity problem. "Labor productivity has been growing at an average rate of only 1.3 percent annually since the start of 2005, compared with 2.8 percent annually in the preceding 10 years," wrote Tyler Cowen in the *New York Times*. Chad Syverson, a professor of economics at the University of Chicago Booth School of Business, says the productivity slowdown has led to the cumulative loss of $2.7 trillion in gross domestic product since the end of 2004. That is how much more output we would have had if the earlier rate of productivity growth had been maintained.

We are today wrestling with a similar challenge presented by the advent of electricity as we attempt to envision the potential applications of digital technology, anticipate the challenges and adjudicate conflicting interests in order to achieve its full potential. We are now caught up in the "gadget phase" of the digital revolution, awaiting the latest new tricks from Silicon Valley. But gadgets are a distraction, not the main story.

The core challenge is using digital technology to boost productivity – like electricity did. How do we quickly translate the creative genius of digital technology into useful applications in industry and commerce? How do we harness the new power and bend it to our will? In sum, how can we grease the skids and accelerate the march of technology to get our economy growing?

For now, there is a great debate underway about just how much impact the new digital technology is having on our economy, and how much more it is likely to have. "The truth is that if you step back from the headlines about the latest gadget, it becomes obvious that we've made much less progress since 1970 – and experienced much less alteration in the fundamentals of life – that almost anyone expected," wrote Paul Krugman in the *New York Times*. "Why?"

Krugman was pondering this issue in his review of a new book, "The Rise and Fall of American Growth, The U.S. Standard of Living Since the Civil War," by Robert J. Gordon, a distinguished macroeconomist and economic historian at Northwestern University. Gordon is a contrarian, arguing against the conventional wisdom that we are in the throes of a great revolution driven by the new technology that will remake our lives like electricity did. Gordon contends that developments in information and communication technology simply do not measure up to the great changes of earlier eras, at least in terms of impact on productivity.

Gordon cites five seminal Great Inventions that powered economic growth from 1870 to 1940: electricity (of course), urban sanitation, chemicals and pharmaceuticals, the internal combustion engine and modern communication. (We need to keep in mind we were making great strides in communication before the digital age – such as telephones, television, radio, fax machines, etc.)

To be sure, Gordon has nothing against technology. To the contrary, he is fascinated by it. He talks about how hard life was for ordinary people before electricity. Before washers and dryers, handling a single load of washing used about 50 gallons of water which some poor housewife had to haul from outside, a well or a creek, 8 to 10 times a day. The women of rural America ironed clothes with heavy irons heated over wood burning stoves. It was laborious work and when the irons got smudged from the fire they had to start all over. Those implements were called "sad irons" and with reason.

But all of that changed almost overnight at least by historical standards. In 1870 there were no homes with electricity and precious few with indoor plumbing or central heating. By 1940, about 40 percent of homes had central heating, 60 percent had indoor flush toilets, 70 percent had running water and 80 percent had electricity. "Except in the rural South," Gordon wrote, "daily life for every American changed beyond recognition between 1870 and 1940." In Gordon's rather dismal view, everything great that has happened since has been part of an echo of that great wave of innovation, and he frankly does not expect us to see anything else like it again, at least not in our lifetimes, he said, because the great inventions cannot be repeated.

Gordon is unimpressed with the Internet revolution, which is the main reason I am so interested in his book. He acknowledges that the Internet breakthrough is highly visible and disruptive, but contends it affects mainly the information, communications and entertainment sectors. He acknowledges the ubiquity of smartphones but sees no direct impact on productivity and does not anticipate one.

There even seems to be a drop-off in introduction of new high tech gadgets. Hayley Tsukayama, reporting from the annual Consumer Electronics Show in Las Vegas for *The Washington Post*, said "Many of the coolest gadgets this year are the same as the coolest gadgets last year – or the year before." The booths are still exciting, he conceded. "It's still easy to be dazzled by the display of drones, 3-D printers, virtual-reality goggles and more 'smart' devices than you could ever hope to catalogue. Upon reflection, however, it's equally easy to feel that you've seen it all before."

Some companies are pushing the concept of the "smart home" but they tend to make it overwhelmingly complex because there are so many compatibility issues to deal with. "No average person wants to figure out whether their favorite calendar software works with their fridge or whether their washing machine and tablet will get along," Tsukayama wrote. "Having to install a different app for each smart appliance in hour home is annoying; it would be nicer if you could manage everything together."

Tsukayama said the high tech companies are too busy pushing their own brands and products when they need to be focused on interoperability and simplicity. "Companies that have long focused on hardware now have to think instead of ecosystems in which everything seamlessly works together to give consumers practical solutions to everyday problems."

Predictably, the Silicon Valley geniuses responsible for much of the digital revolution are more sanguine about the impact of information technology on the economy, insisting the impact simply doesn't show up in traditional measurements of wages and productivity. They point at all of the new things that are free like Facebook, Google, Wikipedia and the like. They insist things are getting better. It just doesn't show up in the data.

Syverson is having none of that. In a recent paper published by the National Bureau of Economic Research, "Challenges to Mismeasurement Explanations for the U.S. Productivity Slowdown," he notes that a slowdown has come to dozens of more advanced economies at about the same time indicating it is a general phenomenon. And countries with smaller tech sectors still have comparable productivity slowdowns, which is not what one would expect if a lot of unmeasured productivity were hiding in the tech industry."

Tyler Cowen, writing in *The New York Times,* contends the tech economy simply isn't big enough to account for the productivity gap. "That gap has caused measured GDP to be about 15 percent lower than it would have been otherwise, yet digital technology industries were only about 7.7 percent of GDP in 2004."

So why is the digital revolution not having a more

pronounced impact on productivity? "Economists and economic historians think they have an answer," Christopher Mims wrote in *The Wall Street Journal.* "To put it bluntly, they say, the problem with the current technological revolution is that, despite multiple Internet booms, we have yet to figure out how to allocate enough capital to information technology and all it enables."

To get the IT revolution up to scale, Mims wrote, "will take more than the markets we have today. We're going to have to transition to the building of public infrastructure and away from the revolution being the domain of private enterprise. It's not enough for Google to roll out high-speed fiber to a handful of cities – we have to recognize that cutting-edge technology yields whole new classes of public goods that we should be pouring money into.

"In sectors from transportation to communications, markets are doing a great job of creative destruction," Mims wrote. "What comes next – what can't happen without a larger, more coordinated efforts – is deployment of technology in a way that benefits everyone.

"The next phase, the one we have yet to enter, is the one in which a technology becomes not merely ubiquitous but also a kind of infrastructure that defines what it means for a country to be developed," Mims said. "It's much more than just rolling out an innovation – dropping computers on desks or building rail across the plains. It's what happens when people learn how to use a technology to make themselves radically more productive."

## Overview

At the Henry Ford Museum in Dearborn, Michigan,

just outside Detroit, there is a marvelous collection of the great innovations of the late 19th and early 20th centuries that denoted this nation's transition into the industrial age. There are many wonders to behold there, like a couple of the huge steam locomotives that once hauled long freight trains across the continent. They are steel monsters. It seems hard to believe normal men created them and operated them. A good engineer knew how much weight he was pulling, how challenging the grade he faced and how to build up steam to achieve maximum power when it was needed. When the locomotives came chugging by, my mother would scurry outside to quickly bring in the fresh washing from the lines before being enveloped by the cloud of smoke from the train.

But the one exhibit that always sticks in my mind consists of huge steam engines, many of them two or three stories high, that powered our early industries before the full power of electricity or internal combustion engine was realized. Looking at them, I could not help but think of dinosaurs – mighty creatures that once dominated the landscape only to be consigned to oblivion. It was the ultimate expression of steam power – a form of energy that even then was nearing the end of its long run. But it did not happen overnight.

We know the technology revolution augurs great progress, but how can we step it up? This challenge is so big that government must take the lead – fostering broad-based collaboration of business, science, academia, entrepreneurs, engineers, government and risk capital to hasten the interconnection of all computer-based systems we depend upon, expanding the Internet of things, making fast Internet access available to everyone, and minimizing silos of proprietary software in favor of universal service and accessibility. In particular, we need to encourage more rapid infusion of advanced technology into our two most stubbornly inefficient sectors – medicine and education – where it is urgently needed.

But the first task is to recognize the fundamental challenge – accelerating the process of integrating advanced technology into our economy to boost productivity. This will be a highly complex undertaking demanding buy-in from many. In a year, we will have a new President. He or she should immediately convene a high level commission representing all affected sectors to develop a comprehensive plan to accelerate the digital revolution, realizing its potential to boost economic growth. Electricity took a century; we cannot afford to wait that long for this.

# CHAPTER SIX

# Saving The Public Schools

*I never let my schooling interfere with my education.*
—*Mark Twain*

As mentioned earlier, publication of "A Nation At Risk" in 1983 sparked a nationwide quest to improve K-12 schools and student performance that has consumed vast sums of money-- with very little results. I took a keen interest in the topic when I was with the U.S. Army because in the Signal Corps we were having a hard time finding recruits who could master the skills sets we required. My new colleague at the U.S. Chamber of Commerce Jeff Joseph had a similar epiphany but he was unable to get senior management very interested in it. "They recognized the problem, but did not see it as a major component of the U.S. Chamber's mission," he told me. "They also did not see it as a means to raise revenues, which is always a priority with trade associations."

But Jeff saw it as crucial to the long-term success of business, not to mention the country, even as I saw it as integral to national security. It was the major reason for our commitment to our Community Learning and Infor-

mation Network (CLIN). Even then, we saw online technology as a critical link in any viable plan to improve the
public schools.

For those who use standardized test scores to keep
track--student performance remains dismal-- as scores nationwide remain far below students in other industrialized
countries. We periodically "dumb down" the tests to miti-
gate the apparent erosion of student achievement. There
are also continuing debates about the utility of charter
schools, the efficacy of teaching to the test (No Child Left
Behind and Race to the Top – both now discredited) and
the use of computers to supplement teacher instruction.
For decades, network news programs and major publications have highlighted interesting and innovative efforts to
transform public education, but nothing has ever stuck and
swept the country.

In the meantime, the chronicle of a failing education
system continues to haunt our society and economy.
About one out of four young Americans lacks a high
school diploma. Even many of those who do graduate lack
the requisite skills to pursue higher education, take responsible jobs in the economy or serve in the military. About
70 percent of eighth graders score below proficiency in
math and reading, and by that time it is too late for many
of them.

We are in fact producing a large cadre of marginally
educated young people with little chance of ever achieving
a decent standard of living, or making significant contributions to society. As recently as the 1960s, the United States
led the world in a variety of important educational
measures. Yet in recent years, the Organization for Economic Cooperation and Development (OECD) ranks the

U.S. 24th or 25th in math and science, in an era when math and science dominate the new economy. The figures for reading are equally dismal. McKinsey Research found that that if the U.S. had been able to close the education gap, our Gross Domestic Product would have been $1.3 to $2.3 trillion higher in 2008.

In the intervening years, many in the high tech world – including several prominent billionaires – have made noisy commitments to improving education, invariably based upon introduction of computer technology to the classroom. They have spent vast sums of money on a variety of experiments that have produced scant results. This experience was not all wasted, however. At least it made clear to Jeff and me that computer technology alone was not the answer to our education dilemma. The real answer is in fostering an upbeat learning environment in the classrooms, and that depends on inspired teachers. Computers have a place, but they are not a panacea.

It occurred to us and many others that one avenue to better schools would be to identify schools that succeed and attempt to replicate what they are doing. What a marvelous idea. And we thought of it ourselves without benefit of consultants or focus groups.

And for sure others shared that idea. From time to time, enterprising journalists spotlight a successful school as a likely model. For example, on February 13, 2012, another "national model" was crowned. That day the *New York Times* published a report on the East Mooresville Intermediate School in Mooresville, North Carolina that appeared to offer a breakthrough of sorts based on issuing Apple laptops to 4th – 12 graders. But though students there demonstrate improved performance based on standardized tests, the Mooresville prototype does not offer a persuasive answer to the national challenge for three basic reasons:

- The school is totally committed to expensive Apple computers and software that necessitated the layoff 65 jobs, including 37 teachers to find the money for an ecosystem that limits students to Apple only products and software;
- Students in K-3 grades are left out of the laptop program- even though 1-2 year olds are routinely playing with laptops, tablets, and educational toys at home;
- The program seems to be designed without a strong vision for replication, and thus difficult (and expensive) to transfer to other systems.

More recently *The Washington Post* featured in its Sunday magazine of March 6, 2016, a school in the Washington suburbs that was performing minor educational miracles. But that was an elite school catering to the children of the upper crust who can afford to pay a steep price to help their kids get a leg up.

Several years before we identified another prototype-- that does not use textbooks and that has produced more impressive results-- over a longer period of time (11 years) -- to all grades K-12. This prototype is not wedded to any one proprietary technology platform and was designed with replication in mind. This prototype is affordable in even the poorest of communities. This prototype is now developing enough "buzz"—that it is poised to break into public view.

I have first-hand knowledge of this school because the late Jack Taub, who provided the creative genius behind CLIN, helped conceive it and was committed to it. Founder of the first online digital network, The Source, Jack was interested in many things, but his most profound interest was in education. He was convinced – and he convinced me – that our public schools are trapped in 19th century technology and methodology. And Taub introduced me to a school and some innovative educators who he believed had the answer.

The Tracy Learning Center (TLC) is in Tracy, California, a bit east of Oakland, a city of 80,000 with a high concentration of migrant farm workers which correlates to a student body where 55 percent receive subsidized lunches. But it also has a fair number of students whose parents work at the world-renowned Lawrence Livermore Labs where brainy people are doing brainy things. So the school population offers a fair mix that reflects the nation's schools overall.

In 1999, Tracy's Superintendent of Schools, a former California Superintendent of the Year, along with a dedicated cadre of reform minded teachers, knew they needed to prepare all students for a completely different world of the future and thus they launched the TLC in an old building in a marginal neighborhood.

This experimental school model was designed to make sense to 21st century students. Moreover, a hierocracy of teacher leadership would empower motivated educators to run the school, with minimal administrative staff, so more resources could be put into the classroom. TLC's design also featured a longer school day (8-4), a longer school year (11 months), no teaching to the test—and-- six figure salaries for its best performing teachers.

So they got to work. Building a school around no textbooks a decade ago required the faculty to develop their own project-based curricula-- with special emphasis on science, math and engineering. So they developed an exciting project based environment that stimulates the inherent curiosity that lives within every child.

This is a critical factor because teaching to the test, the inevitable result of the No Child Left Behind program of the Bush Administration and the Race to the Top program of the Obama Administration was to coerce teachers into doing precisely that – teach to the test. As a result, students develop skills for memorizing and recalling facts for the test. At TLC, the students find, use and learn the same content-standard information, but the learning is done within a context that connects it to its application for solving a problem. This distinction is critical. Students who memorize information for the purpose of testing soon forget the information or are not prepared to use it, even if they can recall it. But at TLC, the students learn and process information within a context and enjoy much greater recall later because of the association. (It is a standard routine of memorization to associate items for recall, creating more than one point of reference in a person's mind.)

Technology is part of the TLC program but not the end-all and be-all of it. Taub liked to speak about an early, failed attempt by General Motors to replace workers with robots. "After several years and several billions dollars," said GM CEO Roger Smith, "we have proved conclusively that we can make defective cars faster." Taub said GM's mistake was to try to use technology to keep doing what they had been doing. But when you do what you did, you get what you got. "Digital textbooks are an example of education's way of automating bad practice," Taub said. "Education has to learn from GM's mistake – forget the old classroom paradigm that worked for an agrarian economy and was modified to work for a manufacturing economy. The global economic, high performance, information rich world our graduates will fact requires a comprehensive redesign of the education system"

Today, many people understand that our education system is deficient, but like GM in the 1980s, they are trying to use technology to make an archaic system work better. What is needed – and what is being demonstrated at TLC – is a new vision of how an educational system can excite young people about learning.

In the TLC discovery and innovation model, students are not given textbooks or subjected to long-winded lectures. Rather:

- A multi-disciplinary problem or project is assigned to a class composed of several small groups of learners. Each project is designed to create a situation where the students will learn a selected set of academic content standards from several disciplines.
- The students' first challenge is to determine what they need to know to solve the problem and decide which students on their team will find which information. Then they must figure out where the information is and go find it.
- The students then determine which information is relevant and valid and which is not.
- The students then share the information with the rest of the team for further validation and analysis.

- Next the students analyze the accumulated information, synthesize it and draw their conclusions.
- Then the students present their conclusions to the class and the instructor.
- The final stage is to engage the entire class in a discussion of the topic itself, followed by a discussion of any differences between the various groups' findings and a discussion of how and where the best information was found, and finally to create a composite solution integrating the best elements of all of the solutions presented.

In sum, rather than focus on teaching information in a discipline silo, the students learn it as a part of a multidisciplinary context – as they will in real life. Equally important, by working in teams they develop a set of learning, working and problem solving skills that prepare them for projects that are not in the textbook, but in the workplaces where they will spend their adult lives.

There is yet another major differentiator in the TLC vision that needs to be recognized and appreciated. Every student is empowered to play and active planning and decision making role in the learning process. The students determine what approach to use in solving the problem. The team decides how to identify and use its resources. In a traditional school setting, the students are passive participants in the process. In a discovery and innovation environment, the students are empowered to take control of their own learning.

The TLC model does include a longer school day and a longer school year – 11 months. The old system was developed in an era when many if not most families were dependent on agriculture. It made sense to have the summer months off so young people could work on family farms. But that era is long gone. In the modern family, both parents are working year round which makes it difficult to keep track of young people released from school and with nothing to do. The country people used to say the devil finds work for idle hands. The TLC vision is predicated on keeping everyone busy. It is a radical concept and one long overdue.

I should add that the TLC students also get a strong dose of practical workplace experience and civil responsibility as part of their curriculum. Each student will do at least two internships at local businesses or agencies before they are graduated, and each student must complete at least 200 hours of community service.

And everyone − students and teachers − love it! The teachers do not complain about the longer days and 11-month school year, in part because they are having fun working with positive students, but also because they are making excellent money − six digits in many cases. Student performances at all grade levels are near the top, with their Primary and Middle Schools scoring well above the threshold required to be listed in California's highest performance category, though Tracy does not "teach to the test." A recent evaluation by the University of Southern California's School of Education ranked Tracy's primary school #3 among all state charter schools. (It should be noted that the two ranked higher are high tuition affairs catering to the elite. Tracy is a working class school.)

The average yearly dropout rate at TLC is less than 1 percent, absenteeism is negligible and teacher turnover is virtually nonexistent. TLC does all this receiving $5500-$7000 per student in public funds, and the school still generates an annual surplus of $500,000 which is reinvested in the system. Best of all, the TLC system was designed to be replicated at other schools around the nation. It could serve as a proven prototype for effective nationwide reform.

At a time when school systems are cutting budgets and laying off teachers to maintain a 9 month calendar year-- that this affordable model which generates a surplus over an 11 month calendar-- is ready for replication.

Over the past several years, we have brought a series of important educators to visit Tracy and see this revolutionary teaching technique in action, and they invariably come away favorably impressed, but we cannot seem to mobilize a movement to replicate it on a broad scale. I have come to realize that the educational reform movement has become an entrenched bureaucracy in and unto itself, spending countless billions every year to no discernible result. A few schools here and there achieve notable results, but they are but drops in the bucket. There are 15,000 school boards out there locked in the past. Most school systems are at last employing digital technology to enhance the educational experience, but as Taub observed many years ago – they are like GM trying to use technology to improve an archaic system.

The TLC vision is surely not the only one proven to work effectively, but it is one of the few I know of that was deliberately designed with replication in mind. The TLC people estimate it will about a year to convert a standard public school to the Tracy system, but the results will be extraordinary. It means enthusiastic, successful students; committed, well paid teachers; and a lower cost for taxpayers. What's not to like?

But it will take money to get the ball rolling, and I and some colleagues may have solved that problem, too. When we were designing CLIN, at the behest of our friends at DARPA, we acquired patents for the online learning technology. Today, many years later, we look around and see any number of schools and for-private enterprises using our technology to provide instruction online. A few years ago, we approached a few of these organizations politely suggesting they should share a small portion of their revenues with us because they were using our patented technology.

Not surprisingly, that approach got us nowhere fast. Initially, I we were reluctant to take them to court because we know lawsuits can be an expensive, time-consuming process with no guarantee of results. "I was ruined but twice in my life," said Voltaire, "once when I lost a lawsuit, and once again when I won one." Having become embroiled in litigation over this matter, we appreciate Voltaire's wisdom more than ever. Once the lawyers get a grip on your issue, the money just begins flying out the door. Every time you think you have the matter settled, you embark upon yet another round of appeals and discovery. Litigation is not for the faint of heart, or the short of pocket.

We have of course endeavored to interest some of the foundations founded by Silicon Valley billionaires that are big on education reform. But we have discovered that once the billionaires sign the checks, the professional foundation people take over and they have their own agendas – namely to keep the money flowing into their pockets indefinitely. They speak only to each other and are not interested in replicating any local experiments in education reform, no matter how successful they may be. As for the billionaires who want to save the schools, they no longer have any control over the foundations they have funded. I have seen this bizarre dynamic in practice – professional foundation managers dismissing opinions of the people who pay their ample salaries.

But we are determined and we do appear to have the law on our side. If we are successful – and one never knows for sure how a jury will rule – we anticipate generating enough funding to seriously fund the Tracy experiment on a large scale. A few years from now, we could be leading a nationwide campaign to replicate the Tracy model everywhere.

We talk a lot about what students should get from school. There is a growing movement, and one I wholeheartedly support, to get our schools and institutions of higher learning more in sync with the workplace. We have tens of thousands of young people graduating from college, laden with huge debts, who cannot find decent jobs. Many of them are tending bar or driving taxis. There is nothing dishonorable about tending bar and driving taxis, but you don't need a college degree to do that kind of work, or the debt that usually goes with a college degree.

And while I am on the subject, I do believe one of the key purposes of education, be it secondary or community college or four year university is to teach young people how to know when someone is speaking nonsense. That is a prerequisite for a viable democracy. The campaign of 2016 strongly suggests our schools at all levels are failing in this basic task.

# CHAPTER SEVEN

# The Dark Side of the Internet

*It's a sin to believe evil of others, but it's rarely a mistake.*
—*H.L. Mencken*

Recent news accounts are detailing how Islamic radicals use the Internet to lure fresh converts into suicide missions. Certain Presidential candidates, most notably Hillary Clinton and Donald Trump, have suggested we should proscribe that sort of activity on the Internet, simply forbidding the radicals to use it to promote acts of random violence.

I have a news flash for Clinton, Trump and anyone else who thinks we can regulate the Internet in such a manner. That horse left the barn a long time ago. There were a few of us way back in the day warning of what could happen, but no one eager to listen. Now our worst fears are coming to reality, and there is in reality very little we can do about it.

Not so long ago there was such a thing as a moral code in this country built essentially upon the core values of a population of people raised with strong religious convictions. Even people who lacked religious faith tended to respect those who had it. But the Internet age has spawned an erosion of standards in the public sphere. There are today few if any restraints on public discourse.

Once upon a time, the Motion Picture Production Code, sometimes known as the Hays Code, proscribed the use of nudity, profanity and other racy material in movies. The major television networks were if anything even more restrictive because they were sending material into people's homes. In the early 50s, there was a serious debate among TV producers whether to permit the representation of a killing in a TV drama. (They finally compromised and permitted the writers to kill a banker – assuming the public would not find that so offensive.)

Nowadays of course you flick on your computer and you can if you wish instantly summon the most extreme forms of pornography to your screen. All of this apparently falls under the rubric of free speech, though there is seldom any speaking involved (or so I am told). I do not believe the Founding Fathers imagined this when they drafted the First Amendment, but then I doubt if they imagined the right of individuals to own machine guns when they drafted the Second Amendment if only because there were no such things back then. Now the restraints are gone and there is precious little we can do about it.

It all began with what was the ARPANET, a computer network created by the Defense Applied Research Projects Agency (DARPA) and used by the U.S. government to share sensitive data in the 1970s. The ARPANET gave way to what we know as the Internet which actually materialized in 1991 when an English programmer introduced the World Wide Web as a place to store information, not merely send and receive it. Up until then the system was mainly used to send e-mails and post notices or articles in forums. But from that moment on, users could create and find web pages for just about anything, and anyone could post information at will.

As the World Wide Web grew by leaps and bounds, users had a problem learning how to navigate it. That led to a series of innovative browsers or search engines that facilitated the search. I have already mentioned what I believe was the first one, The Source, created by my late friend Jack Taub. That morphed into CompuServe and then we had AOL which seemed to own the world for a while and now of course we have Google, Yahoo, and a bunch of others competing for market share.

These modern search engines can predict your search, interpret multi-word inquiries and serve trillions of webpages. For all that, Google and other browsers have a relative miniscule view of everything that's out there – perhaps as little as 1 percent. Search engines work by "crawling" links on a website. If a site owner doesn't wish to be found, it will not include a link to that page. If a web page has no link, it cannot be crawled or indexed. In other words, it is not easy to find.

The World Wide Web has gone where it will at the behest of anyone who can master the rather simple technology. Most people use it for positive purposes, or attempt to, but not everyone. Unlike other path-breaking communication technologies – like the telegraph, telephone, radio, television – the Web is unregulated, a veritable "wild west" where anything goes and no one is in charge. Whether what goes on is good or bad depends on who is doing what and where and to whom.

The anonymity of the Internet has facilitated a new generation of random brutality in which kids torment each other, pornographers fill the cyberspace with smut and anyone and everyone in public life is subjected to the vilest calumny imaginable. Anonymous voices make outrageous charges against you and there is no recourse, no way to identify the source, no means of addressing the deceit or obtaining redress. You can't sue for libel or slander if you don't know who is spreading the dirt and have no way to identify the source. The bullies have unlimited power to wreak havoc and are not the least bit shy about using it. There is no credible defense.

The Chinese are making a herculean effort to control the Internet on their turf, not to protect public morals or discourage aggressive behavior, but rather to protect their oppressive government from exposure and criticism. Under the leadership of President Xi, the Chinese are returning to the dark days of Mao. Indeed, in China, the dark vision of Orwell is coming to pass. A ministry of truth decides what is true and what is false, and anyone who challenges the ministry's ruling gets a visit by jackbooted thugs who haul them away like in Nazi German or Stalin's Soviet Union. You have nowhere to turn for help or support. You simply disappear. The government's abuse of power is not subject to judicial review or scrutiny by an independent news media.

But even the Chinese with their unlimited power and unprincipled willingness to use it cannot stifle the Internet. They have tens of thousands of agents perusing personal e-mails and tweets, identifying malcontents left and right, hauling people off into the black void of their pseudo-judicial system, but they can't keep up with the volume in their own country, never mind the world around them. Even Xi is taking his lumps and apparently his thugs can't protect him. There is something about the Internet that sets people loose.

## The Deep Web

Everyone reading this knows what I am talking about, but few of you really have any concept of the depth and pervasiveness of the Internet or, if you will, the Worldwide Web. There is the Web you and I use often referred to as the "Surface Web" consisting of common sites and search engines such as Google, Yahoo, Facebook, Twitter, Amazon and the estimated 200,000 web sites maintained by businesses, government agencies, trade associations, public service organizations -- the entire gamut of the human experience. Many individuals maintain their own web sites to promote their professional interests or simply out of vanity. Most all of the Surface Web exists within the normal parameters of human interaction, paying at least lip service to legal restraints and our shared values of decency simply because they are not engaged in illegal or disreputable behavior.

But the Surface Web search engines according to some experts only show about 1 percent of what is actually available online. When you find web pages that your search engine cannot access you have entered the Deep Web. It's not as intimidating as it sounds. When you log into your e-mail account, online bank account or Amazon, you are using the Deep Web.

Public information on the Deep Web is about 400 to 550 times larger than the web as commonly understood.

The Deep Web contains 7,500 terabytes of information compared to less than 20 on the Surface Web. The Deep Web is the fastest growing category of new information on the Internet. Deep Web sites tend to be narrower in terms of focused on specific subjects or areas of interest, with deeper content than conventional Surface Web sites. There you will find a cornucopia of hackers, scientists, drug dealers, astronomers, assassins, gamblers, financial consultants, physicists, revolutionaries, terrorists, police, terrorists, sexual deviates, kidnappers, etc.

Very little of the information on the Deep Web is interesting or dangerous or even offensive. It is mostly raw data that has not been analyzed or packaged in any coherent fashion. (As I have said elsewhere in this book and will doubtless say again, raw data is not very useful.) There are blogs and forums galore. There are also, straight out of ancient Rome, underground fighting tournaments in which people fight to the death for the amusement of wealthy patrons who enjoy watching. Humanity has not changed much over the centuries.

The Dark Web, sometimes called "Darknet," is a much smaller section of the Deep Web, where users employ masked IP addresses to intentionally hide web pages from search engines, web page search forms and even standard web browsers. Andy Greenberg of *Wired* says the Dark Web amounts to less than .01 percent of the Deep Web. You need a special web browser to access it. More Dark Web sites in the U.S. use the TOR Network (short for The Onion Router), which is a collection of "volunteer" computer networks that send users' encrypted traffic to multiple servers before pulling up content. In this manner, a user's browsing session is rendered almost untraceable.

Predictably, there are outlaws out there using the Dark Web to pursue illegal activities – such as prostitution, illegal drugs, and fencing stolen articles. One of the more notorious ones was known as Silk Road established in 2011 which was a one-stop shop for illegal drugs, pornography and fake IDs. The key was to use digital currency and ordinary delivery services such as FedEx or the USPS to deliver the goods. The Federal Bureau of Investigation caught up with Silk Road and shut it down, but imitators have popped up in its place. It's like the Whack-a-Mole game. You usually need to have bitcoin in hand to make a deal.

A variety of criminal elements out there are running all manner of scams and extortion over the Internet. In recent years, there have been massive thefts of e-mail addresses, Social Security Numbers and other sensitive data from government agencies, department stores and credit card companies that enables thieves to access individual bank accounts and trust funds. It is the new big business for criminals that offers a maximum amount of opportunity with minimal risk. Law agencies are struggling to keep up with it.

## Cybercrime

A close friend of mine who knows about such things tells me that cybercrime is and for a long time has been a major headache for corporate America, especially international firms. It is fairly routine for companies to report that they receive tens of thousands of electronic messages daily and that the overwhelming majority of them are corrupt in one way or another, usually efforts to breach the company's security walls. Staying ahead of the bogus traffic is a full time around the clock job that consumes huge amounts of the company's resources, but that is the world we live in. They are defending their corporate castles the same way Medieval knights defended their stone castles.

A growing trend in cybercrime is ransomware – a kind of malware that is proliferating in the U.S. and around the world. Hackers worm their way into a computer system, say a bank or a hospital, and render it inaccessible, demanding a payoff to set it free. It begins innocently enough when you click on a link or open an attachment. Often the hackers demand payment in bitcoin which is virtually impossible to trace.

It is usually a lot easier just to pay up than call the cops, which is what many victims do, which in turn encourages more ransomware. In a nine-month period in 2014, the Federal Bureau of Investigation received 1,838 complaints about ransomware and it is estimated that victims forked over more than $23.7 million. The following year the FBI received 2,453 complaints and victims lost $24.1 million.

The crooks are clever. Most of the time the information is vital and the ransom demand is small as in thousands or even hundreds of dollars. But the disruption to a business can be significant, especially if it has not backed up its data, an all too common mistake. In February 2016 the FBI issued an attention-getting alert saluting the growing sophistication of the malware pirates. "The bad guys burrow into a system often months in advance, map out the network and then deploy the ransomware at what they believe to be the most critical assets of the organization," said James Pastore, a former federal prosecutor in New York who worked on a ransomware case involving the Eastern European crime ring called Blackshades. As reported by *The Washington Post,* that case involved authorities in 18 countries working together to make 90 arrests in May 2014.

Perhaps even more infuriating are the malicious hackers who disrupt normal network activities apparently for the sheer joy of causing trouble. Call them cybervandals. A case in point was a recent event in the Washington area when hackers crippled the MedStar Health network. MedStar is a $5 billion health-care provider which operates 10 hospitals and more than 250 outpatient medical facilities in the region. It was an uneven attack hitting some sections harder than others, but that offered scant relief. "In the inpatient units that I'm aware of, everything is off," said Stephan Frum, a union representative for National Nurses United, who has worked closely with MedStar for 15 years. In an interview with *The Washington Post,* Frum said, "The system may be working but is no one can access it, what use is that?"

## Cybersecurity

The U.S. government was hit by more than 77,000 "cyber incidents" such as data thefts or other security breaches in fiscal 2015, a 10 percent increase over the year before, according to a recent White House audit. Part of the bump up is a result of agencies improving their ability to detect incidents.

The White House report defines a cyber incident broadly as "a violation of computer security policies, acceptable use policies, or standard computer practices." The White House said only a small number of the incidents would be considered national security breaches.

That last comment should be taken with a grain of salt. One of our leading experts on cyber security is my friend Robert Gates. When he was appointed Secretary of Defense in 2006 his daily intelligence reports on the cascade of cyberattacks directed at the United States took him by surprise. According to writer Fred Kaplan, author of "Dark Territory, the Secret History of Cyber War," a recent book published by Simon & Schuster, Gates was so stunned by the volume of attempted intrusions into American military networks – his briefings listed dozens, sometimes hundreds every day – that he wrote a memo to the Pentagon's deputy general counsel asking at what point did such a cyberattack constitute an act of war under international law. He got a vague reply about two years later. The fact is we are on terra incognita with this form of threat.

So who is out there doing this? Hostile sovereign powers of course, and also international criminal organizations, the usual run of hackers and terrorist organizations. We are constantly reminded of the huge damage hackers could do if they penetrated and sabotaged our electrical grid. Thus far we have been able to stave off potentially catastrophic attacks on our businesses and government agencies, but the volume of attacks is growing exponentially while we struggle to construct a coherent strategy of defense.

Kaplan reports that in 1997 a secret National Security "Red Team" was instructed to test the defenses protecting the Pentagon's computer networks. The National Military Command Center was hacked in a day. The Defense Department's intelligence directorate was then penetrated with stunning simplicity: a member of the Red Team called claiming to be from the Pentagon IT department and explained that the directorate's password would need to be changed. "The person answering the phone gave him the existing password without hesitation. The Red Team broke in."

The following year the computers at Andrews Air Force base near Washington were penetrated, a hack that swiftly spread to a dozen military installations. The breach, code named Solar Sunrise, was initially traced by investigators to an Internet service provider in the United Arab Emirates, triggering speculation that the operation had originated from Iraq. Yet within days a less dramatic explanation emerged. The culprits were a pair of 16 year old boys in the San Francisco suburbs under the aliases Makaveli and Stimpy.

Mike McConnell, appointed director of national intelligence in 2007, lobbied the National Security Agency as well as other federal departments seeking to impart a greater sense of awareness of the problem. "He would bring the cabinet secretary a copy of a memo," Kaplan wrote. McConnell would hand it over. "You wrote this memo last week. The Chinese hacked it from your computer. We hacked it back from their computer."

China is perhaps the biggest threat. It hacked the Office of Personnel Management making off with the personal information of up to 18 million Americans. Kaplan said China engages in highly-organized commercial cyberespionage and intellectual property theft. The principal antagonist of American business was well known across the senior ranks of the Obama Administration. It was the

Second Bureau of the Third Department of the People's Liberation Army's General Staff, also known as Unit 61398 headquartered near Shanghai. It is only one of unit of a cadre of a cyber-force estimated to be in the tens of thousands.

In January 2013, a Defense Science Board task force released a 138-page report on "the advanced cyber threat." The product of an 18-month study, based on more than 50 briefings from government agencies, military commands and private companies, the report concluded that there was no reliable defense against a resourceful, dedicated cyber attacker.

In several recent exercises and war games that the panel reviewed, Red Teams, using exploits that any skilled hacker could download from the Internet, "invariably" penetrated even the Defense Department's networks, "disrupting or completely beating" the Blue Team.

"The network connectivity that the United States has used to tremendous advantage, economically and militarily, over the past twenty years," the report said, "has made the country more vulnerable than ever to cyberattacks." The problem was basic and inescapable: the computer networks are "built on inherently insecure architectures."

Inherently insecure is a phrase that sticks in my mind. Our vulnerability is built into the system. Verizon issued a report that there were 79,790 verified security breaches in the U.S. in 2014, with about 25 percent more penetrations and 55 percent more data losses than the year before. Cybersecurity industry estimates more than $750 billion in economic value is stolen through global cybercrime and commercial espionage operations each year.

Despite all this, it must be said that cyberespionage is a

two way street and we are as good as it as anyone, probably better. Kaplan talks about Access Operations, an elite unit within the NSA that has developed technologies to penetrate computer networks. "Obscure points of entry were discovered in servers, routers, work stations, handsets, phone switches, even firewalls (which ironically were supposed to keep hackers out), as well as in the software that programmed, and the networks that connected, this equipment," he wrote."

During the Iraq War, NSA equipment and analysts were deployed on the ground in a heavily fortified bunker north of Baghdad to assist with the "surge" to crush insurgent militias and terrorist groups. Captured laptops yielded a vast trove of information that we used to identify al-Qaeda leaders.

In 2009, Defense Secretary Gates created a dedicated Cyber Command and within three years, in the way of the Pentagon, its budget went from $2.7 billion to $7 billion. Cyberattack teams grew from 900 specialists to 4000 and the number is still growing. Perhaps the most ingenious operation code-named Olympic Games, a joint initiative among the NSA, CIA and Israel's cyberwar bureau, Unit 8200, injected the famous Stuxnet malware program into the industrial computer systems of Iran's nuclear facility at Natanz disabling thousands of uranium centrifuges.

But the moral of the story is that we are engaged in a ruthless, relentless cyber war against a host of creative, determined enemies and there can never be a clear cut victory. As stated by the Defense Science Board in 2013, "The network connectivity that the United States has used to tremendous advantage, economically and militarily, over the past 20 years has made the country more vulnerable than ever to cyberattacks."

One sector that is clearly not doing its part is academia. Not one of the top 10 computer science programs in the United States – as ranked by *U.S. News & World Report* in 2015 – requires graduates to take even one cybersecurity course, according to a analysis from security firm CloudPassage. Three of the top 10 programs don't even offer an elective cybersecurity course. And only one of the top 36 programs at the University of Michigan requires students to take a security course to graduate.

It seems plausible that a major reason we find ourselves behind the eight ball in cybersecurity is a lack of education and training at accredited schools. It is possible the students could be learning about cybersecurity in other classes, but it's difficult to get a handle on that. More likely computer science professors perceive security as a hands-on subject rather than the high level abstract concepts they associate with higher education. In cybersecurity, as in so many fields, our universities and other institutions of higher learning are simply out of touch with the real world.

In the Internet, we have created a three-headed monster. The good side is its power to convey useful information and empower people to take control of their own lives and do great things. The downside is its power to facilitate abusive behavior and, as in the case of the Islamic terrorists, lure vulnerable personalities into hideous acts against society. And the dumb side is to amass mountains of raw data that we have neither the resources nor the judgement to manage.

The dark side of the Internet is for better or worse part of our lives from now on. It cannot be effectively regulated or policed without sacrificing the basic freedoms we are committed to preserve. It is what it is. We have to make allowance for it and learn to live with it.

# CHAPTER EIGHT

# Digital Dissonance

*Well, if I called the wrong number,*
*why did you answer the phone?*

—*James Thurber*

I was in a coffee shop watching the crowd of young people around me – each and every one obsessed with his or her cell phone device, busily chatting away or texting, oblivious to the people around them. I had observed some of them come into the shop with others, but as soon as they had made their orders and found a seat, they were into their mobile devices. It was as if they were using those things to avoid actual contact and communication with their friends and associates.

"Americans no longer talk to each other, they entertain each other," Postman wrote in Amusing Ourselves To Death. "They do not exchange ideas, they exchange images. They do not argue propositions; they argue with good looks, celebrities and commercials."

There is a current car commercial on network TV featuring some "real life" people who are asked to give up their smartphones which are then tossed into a shredder and ground to pieces. The panic that erupts is truly disconcerting. Even when assured that fake smartphones were substituted for the real ones, the subjects remain petrified. It is as if the very thought of having to get along without their precious digital devices for a few moments is mortifying in the extreme.

Recent news reports show a spike in deaths and injuries among pedestrians. The death rate is apparently at an all-time high and the main cause is preoccupation with smartphones. People are walking down the street focused on smartphone conversations or busily texting, oblivious to the world around them. Sometimes they are conversing through ear plugs which makes it difficult if not impossible for them to hear warning sounds such as horns, sirens or approaching buses and trucks until they are mashed flat.

There is to me something disturbing about the national obsession with these devices. They have become like a powerful narcotic only we have no Drug Enforcement Administration working on this affliction, cracking down on pushers, nor do we have a medical infrastructure set up to help addicts beat their habits. Smartphone addiction is perfectly legal even though it is clearly as addictive and potentially as dangerous as heroin or cocaine. It affects all classes of people from all ethnic groups and across all income levels. It has slipped up on us without discussion or conscious anticipation. It is changing the way we live – the way we communicate with each other, the way we relate to each other, and the way we participate in society.

Look around you, especially at the young people. From dawn to dusk their hands and minds are preoccupied texting, e-mailing, tweeting, watching videos, sending Facebook messages, and playing computer games. The research firm eMarketer says Americans spend an average of five and a half hours a day with digital media, more than half of that time on mobile devices. Among some groups, presumably the younger set, the numbers are much higher.

For example, in one survey, female students at Baylor University in Waco, Texas, reported using their cell phones on average 10 hours a day. Three quarters of 18-24 year olds say they reach for their phones first thing in the morning. According to a study in the United Kingdom, people check their smartphones 221 times a day, or once every 4.3 minutes.

You think this cannot be true? Look around you on the street, in the grocery store, on the Metro everywhere you go. Are you not one of the addicts? Then you are an outlier, a loner, a misanthrope – one of us old fogies. And what could all of those digital addicts with those things glued to their ears possibly be doing 10 hours a day on their smartphones that is in any way useful in terms of career development or building relationships. Frankly, some of those people I see on the street or in restaurants strike me as dull as dishwater. I cannot imagine they have anything faintly interesting to say to anyone, but they are busy saying it anyway.

Jacob Weisberg, writing in *The New York Review of Books (Feb. 25, 2016),* says "our transformation into device people has happened with unprecedented suddenness." Indeed, it has. As Weisberg points out, the first touchscreen operated IPhones came online in 2007, followed by the first Android phones the following year. "Smartphones went from 10 percent to 40 percent market penetration faster than any other consumer technology in history," he said. We had no advance warning, no time to prepare, and now we are unaware of what has happened to us and its implications for our future and that of our nation, not to mention civilization. We might have handled it better had we known it was coming and had time to prepare for it, but it just happened all of a sudden and here we are.

In my lifetime as a military officer in the Signal Corps, I was attuned to the rapid changes in communications technology and especially sensitive to the impact of technology on the behavior of our troops, especially in terms of their attention spans. The people we have managing security of our nuclear weapons, for example, are subject to long periods of isolation and boredom in remote locations. Even in my day it was always a challenge finding ways to keep them alert and fully occupied and I retired well before the onslaught of smartphones.

There is a fundamental flaw in all this online chatting and texting. To truly communicate with someone, you need more than written words or even spoken words. People rarely say exactly what they mean. Most of us are not every articulate to begin with and even when our thoughts are clear they are modified by mood and circumstance. To really communicate with someone you need to look him or her in the eye and get a sense of the mood and conviction. Under the best of circumstances, there is always the risk of miscommunication and resulting confusion and misunderstanding. It do not believe very much of all this chatting and texting is fomenting true communication. Indeed, I believe much of this activity is an escape from communication simply because few of us really have that much to say.

I have recently read two books by a perceptive observer of this changing scene named Sherry Turkle, "Reclaiming Conversation: The Power Of Talk In The Digital Age," and "Alone Together: Why We Expect More From Technology And Less From Each Other," that shed light – a scary light in my opinion – on what is happening to our society. Turkle contends, and I fully agree, that the smartphone revolution is having an adverse impact on normal personal relationships – among family, work colleagues and friends. On the surface, the new technology greatly enhances the ability of people to communicate with each other. But beneath the surface it offers a more sinister option, that of opting out and going alone. A Pew survey of 2015 found that nearly half of 18-29 year olds use their cell phones to "avoid others around you."

Turkle hits the nail on the head. She writes about parents who are distracted and unable to pay close attention to their children; children who are frustrated because they cannot have their parents' undivided attention; gatherings where the people who are really there must compete with invisible participants; classrooms where professors look out upon a sea of disengaged multitaskers glued to their laptops or smartphones; and a dating culture in which infinite choice undermines the ability to make emotional commitments. To that I would add professional and business and professional meetings where participants have their heads into their smartphones ignoring the conversation around them. It can be truly difficult communicating complex points of view even when you have Power Point and someone's undivided attention. When you are sharing their attention with others via digital technology the difficulty is magnified many times. You might as well stay home.

The ubiquitous digital devices not only deter effective communication among people, it disturbs one's inner communication. When I was a young colonel attending the U.S. Army War College in Carlisle, Pennsylvania, I encountered some extraordinary stress due to a tragedy in my family that challenged my inner coping mechanisms. Then as now I relied on my religious faith to sustain me in the dark hours, but while at the War College I discovered something else – transcendental meditation (TM) which is a methodology for clearing your mind and achieving an advanced state of relaxation. (I hasten to add that TM is not a religion or a substitute for religious faith.) The key to achieving that relaxed state is to cut yourself off from the world if only for a few moments and to look inward. You cannot do that if you have a smartphone on your belt constantly sending you calls and texts. To those who seek relief from daily stress in TM, and I highly recommend it, your first assignment is to leave that smartphone or whatever digital device you carry at home and forget about it. It is a distraction and a crutch that you do not need. You first task is to disconnect with the Internet before you can connect with your inner self.

These days I hear and read many complaints that the new technology is not producing gains in productivity. I believe I could make a plausible case that the new technology is in many instances an absolute detriment to productivity and is in fact taking us backwards. For my part, it is less of a tool than an opiate.

Weisberg writes that some young people prefer electronic communication because it makes it much easier to avoid confrontations over difficult subjects, whether in family relationships or other human interactions. I would argue that this is enabling people to avoid a key aspect of communication, acquiring understanding of complex problems and development of positive relationships. Sometimes we must have confrontations – whether with family, friends or colleagues -- to bring everything out in the open and clear the air. There is no other way. To flee from them is to flee from life itself.

In some areas, such as business or the military, such flight also is to invite catastrophe. There are areas of human endeavor where we simply must have brutally clear communications regardless of whether someone's feelings are hurt. To the extent that the digital technology enables people to avoid such unpleasantness, it is a real and present hazard.

In the meantime, we are daily confronted and annoyed by misuse and abuse of digital technology by people who have been catapulted into the new digital age without developing basic social courtesies that generally apply to other, more settled modes of communication. You would think that most everyone would know to turn the smartphone off when entering a church, a theater, a funeral, a wedding, or an important meeting of any kind – but you would be wrong. The same people who program their devices to make all manner of disruptive sounds when a call or text comes in seem blissfully unaware that such calls might come at inconvenient moments and disrupt sensitive situations.

## Social Media

The late Andy Warhol famously said that in the future everyone would be famous for 15 minutes. The future is here and it's actually more like two or three minutes. The attention spans of the younger generation cannot accommodate 15 minute blocks of time. Their fledgling minds have been hijacked by digital technology, broken down into tiny blocks of semi-thoughts on Facebook and Twitter, rendered incapable of critical analysis.

Perhaps the most persistent and annoying example of the digital culture is the so-called selfie – when people using smart phones hold them away from their bodies and take photos of themselves and whoever they happen to be standing next to. This is pure narcissism at work but it is reflective of our times. I have read that media stars – movies, TV, music, even politicians – no longer are asked for autographs but rather to cooperate with a selfie. One actor lamented that people come up to him and without asking permission drop one arm over his shoulder and take the photo with the other hand. All he can do is grin and bear it.

Actually, the first known selfie was taken by a guy named Robert Cornelius way back in 1839, but thank God it didn't catch on. The first use of the term in any paper or electronic medium appeared in an Australian Internet forum on September 13, 2002. But of course the real power of the selfie is the ability to post it on the Internet, which first happened in Australia the year before (though the photo in play was taken with a disposable camera, not a cell phone.) A recent poll commissioned by smartphone and camera maker Samsung found that selfies make up 30 percent of the photos taken by people aged 18-14.

All of which brings us to Facebook. A friend of mine

who shares my skepticism of social media said Facebook is a place where old people (like us) post photos of their latest grandchildren. I do not do that, for which my grandchildren will no doubt someday be grateful. But lots of people do post their photos, including selfies, on Facebook. This too is a reflection of the tendency of digital media to drive wedges between people. A 2013 study of Facebook users found that posting photos of oneself correlates with lower levels of social support from and intimacy with Facebook friends. The lead author of the study concluded that "those who frequently post photographs on Facebook risk damaging real-life relationships."

That is a scary thought because one hell of a lot of people are posting photos and other messages on Facebook. It is now the sixth most valuable public company in the world, with a market value of about $325 billion. That's billion with a B. Facebook claims nearly 1.6 billion monthly users for its social network. About one billion people, which is nearly a third of humanity with access to the Internet, log onto Facebook every day. I am not one of them, but perhaps you are. Posting photos of your grandchildren.

Facebook now takes up more than 20 percent of the Internet time we Americans spend on mobile devices, compared to only 11 percent for Google and YouTube combined. Facebook now probably has more data on users than any other company in history, which has enabled it to become a powerhouse in advertising. Its revenues have more than doubled in two years to $18 billion in 2015.

But the truly insidious aspect of Facebook is the way it channels people into mini-communities of the like-minded. Once you log onto Facebook and register an opinion, interest or preference, the system automatically puts you in touch with other people who share your interests and opinions. As Frank Bruni wrote in *The New York Times* on May 22, 2016, "we're sorting ourselves out with a chilling ruthless efficiency. We've surrendered universal points of reference. We've lost common ground." The end result is that we sever communications with people who disagree with us and who offer contrary opinions. "It's not about some sorcerer's algorithm," Bruni wrote. "It's about tribalism that has existed for as long as humankind has and is now rooted in the fertile soil of the Internet, which is coaxing it toward and full and insidious flower."

Everyone keeps waiting for Facebook to fold overnight like other Internet flashes in the pan, but thus far Facebook seems to have unusual staying power. Part of that staying power is built on acquisitiveness. Facebook has paid big bucks for other companies in the social media field like Instagram and WhatsApp, messaging services that enable users communicate face to face, and Oculus which does something in virtual reality. In 2015, Facebook added 200 million users. Like Ol' Man River, Facebook just keeps rolling along seemingly powered by its own momentum. As with most forms of digital communication, its primary purpose seems to be to enable people to avoid face-to-face contact with each other.

For a while there it seemed like Twitter was going to be the big kid on the block for reasons indecipherable to me it has withered as Facebook has exploded. Many prominent people – including politicians and media stars – use Twitter to great effect. But my friends tell me (sorry, I don't tweet despite the title of my last book) that Twitter is mostly a forum for people trying to sell things. "I post something interesting every day, like an old quote or a snappy bon mot," said one. "I have almost 500 followers. But when I return the favor and look at their tweets, I see pitches to sell things. I do not know why I would spend my time perusing Twitter to see commercials." Well, perhaps as with Facebook he could use it to avoid face-to-face communication with other human beings.

As of this writing, I am not certain where the new Apple Watch fits into all this. It is a device, not a program, so perhaps it will simply provide another mechanism for posting your photos on Facebook, though presumably you will still need a smartphone to take the photo. We may soon reach a time when just everyone has to have an Apple Watch, just like everyone today has to have a smartphone. And of course you can always use it to check the time.

Overall the shift to social media using digital technology is changing the way people behave toward each other, not necessarily for the better, and undermining their grasp of the world around them. The rise of digital technology corresponds to the decline of newspapers, news magazines and other traditional news sources. The young people insist they get their news from the Internet. In fact one can get the news from the Internet – if you know what to look for. To me reading a newspaper has always been an adventure into the unknown. You never know what you will find there. You encounter unfamiliar viewpoints. And just as the rise of digital technology corresponds to the decline of newsprint, so does the decline of newsprint correspond to the erosion of our civic spirit and tolerance of diversity.

Writing before the digital age, during the late stages of the television era, Postman underscored the vital importance of the printed word to human society. "In a culture dominated by print, public discourse tends to be characterized by a coherent, orderly arrangement of facts and ideas," he said. "The public for whom it is intended is generally competent to manage such discourse."

The new digital technology does not seem to be contributing to productivity and it is not fostering closer human relationships. There seems to be something about digital technology that suppresses the normal human consciousness that we have evolved over the years to generally govern our behavior around other people. If we are to abandon time-honored ways, we should at least have some viable idea of what we will replace them with. We are wandering into an unknown land adrift from our core values. This is not progress, this is regression.

# CHAPTER NINE

# Technology In Medicine

*The desire to take medicine is perhaps the greatest feature which distinguishes man from animals.*
—*Dr. William Osler*

Earlier I made reference to the lengthy time it took for our ancestors to make full use of the great advance in uses of electricity. There were prolonged battles over the technology as it evolved, endless patent lawsuits and intense competition among industrialists seeking to gain the greatest advantage from it. The juice was a long time coming to rural America because the challenge of hanging wires many miles to serve only a few customers was a deal killer for the private sector electric companies focused on profits. The government had to step in to get the wires into the mountains, as with the Tennessee Valley Authority, and to the farms. A promising young Congressman named Lyndon Johnson made his reputation by getting electric service into the Texas hill country.

The same complex process attended the evolution of trains as competing interests championed a variety of steam engines and even the proper gauge of track. In the late 1800s, a cadre of railroad tycoons (and tycoon wanna-bes) built railroads all over the place without sensible calculation of how many were necessary or supportable by the potential traffic. Their only idea was to drive their competitors out of business and establish a monopoly. They would often lay competing tracks to the same major urban areas when any sensible person could see that one was more than enough. They fought it out until some of the companies went under. This led to major bankruptcies that threw the national economy into a tailspin more than once. It was chaos for a long time. And when individual companies finally did achieve dominance in specific areas they usually celebrated by imposing burdensome rates for carrying cargo and passengers that worked major hardship on the public. The mess did not begin to get sorted out until government intervened with regulation to protect the public interest.

Today we are seeing much the same thing in the medical industry as physicians, research institutes and hospitals embrace marvelous advances enabled by digital technology that would have been beyond imagining only a few decades ago. We have come a long way since ignorant doctors practiced bleeding of patients such as likely killed President George Washington and ignored evolving knowledge of antisepsis which stubbornness killed President James A. Garfield, one of the most promising chief executives ever to grace the White House. Various doctors took turns probing the bullet wound in President Garfield without bothering to wash their hands. The wound itself would not have killed him. It was the infection caused by ignorant doctors who should have known better. The knowledge was available, but they chose to ignore it.

Today our hospitals, doctors' offices, research institutes and other medical facilities have come a long way. They have at their disposal marvelous new machines that can see inside the body and diagnose a variety of ailments enabling application of appropriate remedies. I am one of those who has benefitted big time from the new knowledge. Long ago I had two major heart surgeries rectifying problems that would have been fatal even a few years before. Today I am in my mid-eighties savoring life to its fullest – playing golf, flirting with the ladies, singing along with patient pianists and writing books. I tip my hat to the wonderful advances of medicine which prolong both life and health for me and millions of my fellow citizens.

At the same time, I see many correlations between the rapid advance of medicine and the expansion of railroads. Today countless companies develop new medical technologies that they jealously guard and protect with patents and exclusive access. You may have tests and procedures at one hospital that are mystifying to other medical practitioners across town who are committed to other systems and procedures. Because of proprietary interests, the medical treatments you receive at one medical facility are not necessarily transferrable to other medical facilities.

## Big Business

Like the railroads I spoke of, medical care has become a big business and as such it evokes highly competitive enterprises seeking lucrative markets. The U.S. medical technology industry is projected to grow at an impressive 6.3 percent Compound Annual Growth Rate (CAGR) to

$156 billion by 2018 driven by significant capital investment generated by two key factors – lower reimbursement risk and an expected increase in patients requiring care under the Affordable Care Act (Obamacare). Implantable devices, including reconstructive joint replacements, spinal implants, cardiovascular implants, dental implants, and intraocular lens and breast implants are expected to grow at a CAGR of 8 percent to nearly $74 billion by 2018.

A growing number of original equipment manufacturers (OEMs) of medical devices are shifting focus to innovation and design rather than manufacturing, depending on outsourcing to supply the goods. The medical device outsourcing industry in and of itself is a large market of about $10 billion and is growing faster than the underlying medical device market.

There will likely be dramatic changes in the nation's health care system within the next few years that will have far-ranging implications for everyone. At present, the U.S. is spending 17 percent of its Gross Domestic Product on health care. In contrast, the other industrialized nations are spending around 10-11 percent, or about 50 percent less. Meanwhile, by most standard indices the typical U.S. citizen is no better off medically than those of other industrialized nations. In fact we are in many ways worse off though that may be primarily due to other factors such as drug abuse, gun violence and others.

The cost of medical care is soaring as the Baby Boom generation enters the prime retirement years of 65-85 when their demands on the health care system will accelerate. Today, the typical family of four pays about $18,000 a year for health insurance. Many to not grasp the significance of the cost because it is borne at least in part by their employers. In fact, many employers – chief executives of corporations – do not realize what a major drag it is on their profitability and competitiveness.

The U.S. is the only major industrial nation that built its health care system on the employment connection. This was in fact a happenstance that occurred during World War II when companies needed labor badly but were not allowed to offer higher wages to attract them. (During WWII, the government closely regulated wages and prices.) Thus employers hit on the idea of offering health insurance as an extra non-monetary bonus. It seemed like a good idea at the time, but it gradually became the norm.

Unfortunately, today millions of Americans are unemployed, partially employed or work for employers that do not provide health insurance. About one out of six Americans is uninsured, a huge number of people. They cannot afford to pay $18,000 a year for health insurance which is the going average for a family of four.

This is not to suggest the unemployed do not have access to health care. Under federal legislation enacted in 1985, no doctor or hospital can deny treatment to anyone who needs it. The result is that poor people generally receive little or no basic medical care until they are in dire straits at which time they go to a hospital emergency room. It would be difficult to imagine a more cost inefficient system than this one. People who cannot afford routine care wait until their condition is extreme and then appear at the most expensive medical station known to mankind – the hospital emergency room. Thus, hospitals charge people with insurance an added premium to cover the emergency room overages.

There is a general consensus among health care experts and politicians that the current slapdash system is not sustainable in the long term. With or without a major shift in our approach to health care delivery, the American people are facing profound changes in the way they obtain and pay for health care – which will have a profound impact on daily life in myriad ways. On a basic level, there will be a sea change in our attitude toward health care as we embrace the concept of universal health insurance, not necessarily because there is a consensus that everyone deserves health insurance, but rather because we cannot afford to continue the existing system of providing free health treatments to everyone in hospital emergency rooms.

Regardless of what happens or doesn't happen to the Affordable Care Act (Obamacare), the Federal Government cannot and will not ignore the soaring cost of health care driven primarily by a growing population of the aged as well as dramatic breakthroughs in medical technology, including new generations of pharmaceuticals. Already, the budgetary pressures are fostering disarray in Congress and are creeping into the political debates. There is simply no way to ignore it.

This is a uniquely American dilemma. Most of the other industrialized nations have in place national health care systems through which everyone is insured, regardless of income or employment status. As early as the late 1940s, the Truman Administration made a serious push to adopt such a national system in this country, but it was resisted and eventually killed by Congress, under pressure from the medical profession. Then as now, there was an abiding lack of confidence in the ability of the Federal Government to handle anything so complex and personal.

The death of national health care paved the way for

further expansion of the employer-based health care system. But now the aftereffects of the great recession and the impact of the digital revolution are exposing vast gaps in the traditional employer-based health insurance system. We cannot indefinitely sustain a system in which one out of six Americans has no health insurance and must depend on emergency rooms for treatment.

Obamacare does almost nothing to control costs directly, except that getting millions of the uninsured to enroll in health care plans will help. Even if they pay less than what their care actually costs, at least they will be paying something into the system and will not be relying on emergency room treatments. Also, assuming they receive basic medical care, there will be far fewer emergency room visits which are often the result of a lack of basic care.

Obamacare is a highly volatile topic because it was largely a one-party creation and lacked bi-partisan support. But also it flies in the face of a general conservative opposition to expansion of the federal government. These are legitimate issues for political debate, but the challenge of bringing health care costs under control and extending health insurance to all citizens remains to be solved.

In reality, the system itself, partly prodded by the federal government, but also by normal economic forces, is shifting in subtle ways that are radically changing the way health care is delivered and paid for, and which eventually will bring everyone into the system.

## Accountable Care Organizations

Like it or not, accountable care organizations (ACO) are the wave of the future as the system moves away from the traditional fees for service to a more comprehensive system that emphasizes results. An ACO is a healthcare organization characterized by a payment and care delivery model that seeks to tie provider reimbursements to quality metrics and reductions in the total cost of care for an assigned population of patients. A group of coordinated health care providers forms an ACO, which then provides care to a group of patients. The ACO may use a range of payment models (capitation, fee for service) with asymmetric or symmetric shared savings, etc.). The ACO is accountable to the patients and the third-party payer for the quality, appropriateness and efficiency of the health care provided. According to the Centers for Medicare and Medicaid Services (CMS), an ACO is "an organization of health care providers that agrees to be accountable for the quality, cost, and overall care of Medicare beneficiaries who are enrolled in the traditional fee-for-service program who are assigned to it." ACOs are:

- Provider-led organizations with a strong base of primary care that are collectively accountable for quality and total per capita costs across the full continuum of care for a population of patients;
- Payments linked to quality improvements that also reduce overall costs: and,
- Reliable and progressively more sophisticated performance measurement, to support improvement and provide confidence that savings are achieved through improvements in care.

This new approach to health care delivery is not yet enshrined in law, but Medicare is taking the lead and Medicare is the 700-pound gorilla in health care. The pattern set by Medicare will be followed by major insurers. Already, hospital consolidations are underway because the hospitals need to control the ACOs from beginning to end. Some experts estimate that 80 percent of all doctors will be working for hospitals by the end of this decade. Hospitals will be paid based on quality of results defined largely by the level of re-admissions and infections. There will be tremendous pressure for cost containment across the board. Medical devices that respond to this pressure will do well; those that do not will go by the board. But overall, the price pressure and emerging technologies will translate into use of fewer medical devices, and hence less use of plastics.

## More Digital Technology On The Way

As inefficient as our medical system is, we remain committed to research into ever more effective ways of treating medical conditions. Some of the research is funded by government, some by academia and some by for-profit companies. In general, this is a good thing. But the research plows ahead indifferent to its impact on rising medical costs.

A couple of years ago I had the pleasure of meeting Paul Allen, co-founder of Microsoft, who has a pile of money at his disposal and only spends a few hundred million or so on that huge luxurious yacht he maintains. I was speaking to him about education, a cause he is very interested in. He and other high tech billionaires have pumped huge sums of money into education with scant results. Computers are tools, not solutions. What our schools need are better teachers and they are hard to find in part because the system overall simply does not encourage and reward superior teaching. Too many schools systems are being run on barebones budgets that can barely keep the heat on, much less reward superior teachers.

But Allen also is interested in health care. I read recently that he has put up $100 million to fund research at the "frontiers of bioscience" that could have profound implications for health care. Allen said the 10 year undertaking grew out of a realization that the biological sciences are at a critical point in history with technology now looking at a quantitative leap in understanding of the human body. New tools, make possible by our advances in digital technology, can manipulate DNA. Next generation microscopes can measure and create images of the tiniest parts of living systems, and super-powerful computers are able to make sense of massive amounts of data.

"What I believe is that this is potentially a game-changer for our understanding of complex biological systems," Allen told *The Washington Post*. He is determined to facilitate a more interdisciplinary approach by giving scientists with out-of-date ideas the equipment, staff and connections to counterparts in math, engineering, physical sciences and computer science – so their work can reach its full potential.

To be sure, Allen's commitment represents a tiny sliver of the $15 billion the National Institutes of Health has doled out in recent years for biomedical research. Even so the Allen awards will be unusually large and unrestricted. James Collins, a bioengineering professor as he Massachusetts Institute of Technology, will use his grant to engineer organisms that could help to trap and kill bacteria resistant to traditional antibiotics. That would be a major break-through. "What we know about what antibiotics do to the body is remarkably incomplete," Collins said.

Allen is just one of the growing number of tech industry tycoons using their big bucks to advance medical science. Facebook's Mark Zuckerberg, Google's Sergey Brin and others have teamed up to create the Breakthrough Prizes which are awarded each year to scientists in various fields – now known as the Silicon Valley Nobels. Napster's Sean Parker has funded an institute that aims to cure allergies, eBay's Pierre Omidyar is backing research on resilience, and PayPal's Peter Thiel has taken aim on "breakout ideas" in science which could mean anything.

I have read recently of significant progress in the field of immunotherapy that enables our bodies to kill cancer cells, potentially superseding the current therapies of surgery and chemotherapy. This research is greatly aided by the power of digital technology.

**The Bureaucratic Morass**

All of this is exciting and inspiring, but one of the most obvious targets of the new technology – and one that got a lot of discussion when the promise of technology was first coming into view – was that of the bureaucratic morass that engulfs our health care delivery system. One would think that these masters of the universe could apply their collective genius to figuring out creative ways to manage the paperwork that engulfs medical care. Is there anyone among us who has not been driven to distraction by the myriad challenges of managing their health insurance, filling out the right forms, arguing with clerks, trying to figure out what is covered and what is not?

It was once thought that modern information technology would translate into online medical records that would relieve doctors and hospitals of the eternal paper chase trying to find a patient's records and glean the important information from them quickly. If there is any aspect of our medical challenge that should be amenable to information technology, it is surely this paperwork nightmare that costs a fortune, leads to inordinate delays, and fosters endless mistakes that are often fatal.

But this challenge has thus far proven insurmountable, and the example of the railroads offers a clue why. What we need is a common system of shared data bases, but that is not possible when the field is cluttered with thousands of competing health insurance companies, health care providers and proprietary technologies that incompatible with others. Many physicians are striving to shift to digital records but it is a massive undertaking because they have years of paper records for millions of patients. They haven't the time or resources to convert them all into digital format. And as for the competing propriety systems, we are back in the age of the railroads, struggling to forge a unified, coherent system amid the cacophony. Digital breakthroughs can provide wonderful treatments and cures for disease, but they have yet to offer a solution to this bureaucratic maze.

# CHAPTER TEN

# The Inequity Enigma

*I've been rich and I've been poor. Rich is better.*
*—Sophie Tucker*

There is no reasonable doubt that the distribution of wealth in our country is profoundly unequal and becoming more so by the day. Figures don't lie but liars will figure, so one has a choice of which data to accept about wealth inequity, but regardless of which data you prefer, the gap is huge and growing. As of 2010, the richest 1 percent of the population had 34 percent of the accumulated wealth; the top 0.1 percent had about 15 percent. And the disparity is getting more pronounced. Since the Great Recession, the top 1 percent has enjoyed 95 percent of the income growth from 2009 to 2012 if capital gains are included.

The top 10 percent today accounts for 48 percent of national income; the top 1 percent makes almost 20 percent and the top 0.1 percent nearly 9 percent. In his widely read book "Capital in the Twenty-First Century," author Thomas Piketty says that wage inequality in the United States "is probably higher than in any other society as any time in the past, anywhere in the world."

It is incorrect to state, as Donald Trump likes to do, that this is a "poor country now." In 1950 our Gross Domestic Product (GDP) was about $2.2 trillion as measured by inflation adjusted "2009 dollars." In 2015, our GDP was $16.3 trillion – more than seven times greater. Even accounting for a doubling of population in that period that still leaves today's economy four times larger than its 1950 counterpart. The problem is not aggregate wealth; the problem is distribution of wealth. A disproportionate share of our national wealth is going to the people at the top who need it least. Working people are barely holding their own.

This stark contrast must be disturbing to any and all who are concerned about our nation's social and economic stability. History teaches us that income gaps of this magnitude are often a precursor to chaos and revolution. Oddly, the American people have been slow to pick up on this emerging divide. In a 2011 paper, Michael Norton and Dan Ariely reported on a survey of more than 5000 Americans about their perception of the wealth gap. The typical citizen they found believed that the richest fifth of the population owned 59 percent of the wealth and the bottom 40 percent owned 9 percent. In reality the top 20 percent of U.S. households owns more than 84 percent of the wealth and the bottom 40 percent a meager 0.3 percent. The Walton family, primary owners of WalMart, have more wealth than 42 percent of American families combined.

In a more recent study, Norton and Sorapop Kiatpongsan asked people how much they thought CEOs and workers should earn. The median American estimated that the CEO to worker pay ratio was 30 to one, and should be more like seven to one. In reality it is more like 354 to one. "In sum," the researchers concluded, "respondents underestimate actual pay gaps, and their ideal pay gaps are even further from reality than those underestimates."

For a variety of reasons, this growing disparity of income has not to date become a major political issue, in part because most Americans seem committed to the traditional American dream. Pew Research says most Americans believe the economic system unfairly favors the wealthy, but 60 percent believe that most people can make it if they are willing to work hard. Senator Maro Rubio (R-FL) said America has "never been a nation of haves and have-nots. We are a nation of haves and soon-to-haves, of people who have made it and people who will make it."

Then again, the collapse of Rubio's campaign for the White House may reflect more than his penchant for foot in mouth disease, a common failing among politicians. It may be that the American people are beginning to lose that latent optimism about their economic prospects. Reality is beginning to sink in. Today, in a time when the ranks of billionaires continue to swell from year to year, the inflation-adjusted median wage of American males is lower than in 1969, and median household income is lower than when this century began. "If the growth rate since 1970 had matched that of 1920 to 1970," columnist George Will wrote, "instead of being one third of it, per capital gross domestic product in 2014 would have been $97,300 instead of $50,600."

The reality is that the American middle class is being hollowed out. The middle class jobs that once provided living wages and benefits, enabling generations to live well and assume their children would do even better, are disappearing from the economic landscape. Many people are doing well and a handful are doing extremely well, but the streets are filled with desperate people living from paycheck to paycheck, often holding down more than one minimal wage job while living in fear of being laid off on any given day. The overall economy is chugging along with marginal growth of about 2 percent annually, and vast fortunes are being made, but the money is not trickling down to those at the bottom of the pyramid. The tide may be rising but many boats are stuck in the mud.

There is an important distinction between income and wealth. Income refers to how much money you have coming in on a weekly or yearly basis. Wealth refers to the accumulation of assets over time. Many of this nation's wealthiest people have no paying jobs, but rather live on income generated by their fortunes or business holdings. A substantial number of successful CEOs, thespians, athletes and more than a few criminals live off of their accumulated wealth. Presumably they are counted by the Bureau of Labor Statistics (BLS) as "discouraged workers" in that they are no longer seeking employment. This category, however interesting, does not really impact the overall picture of a nation devolving into haves and have-nots while the traditional middle is hollowing out.

## The Pervasive Impact of Technology

The question is – why is this happening? There are many reasons, but a core reason, perhaps the most decisive reason, is the pervasive influence of information technology, the digital revolution whose marvelous power is remaking virtually every phase of how we live and work. It is skewering the economic environment at the expense of the many to the advantage of a few.

There are without question certain parallels to what happened in this country in the late 1800s when steam power and electricity remade the American workplace, fostering greater efficiency and rendering obsolete many traditional job skills. Economists offer bland statements about how this all worked out for the better as the nation became more wealth and the displaced workers found new jobs in a more advanced economy.

What they leave out is the many years of social unrest – strikes, bombings, killings – that invariably attend such dramatic transformation. And many of those displaced workers never did find viable new careers. They were essentially pushed aside, condemned to lives of poverty and despair, and this at a time when there was essentially no social net to aid their families. Later generations benefitted mightily from the industrial revolution, and we are still benefitting from it, but the transition was hard for most working people.

It was not at all hard for the handful of people at the top of the pyramid, many of whom made vast fortunes creating family dynasties that still loom large across the American landscape – Rockefeller, Astor, Vanderbilt, Carnegie, Huntington – in the so-called Gilded Age. They amassed vast fortunes in the hundreds of millions – tens of billions in today's currency – in a time when there was no income tax. The income disparity was so extreme that the American people in 1913 demanded and got the 16th amendment to the Constitution authorizing an income tax (which the Supreme Court had earlier ruled unconstitutional.). The idea was that only the very wealthy would have to pay an income tax, and for many years that was the case. (Just remember next time you're complaining about your taxes that our ancestors insisted on an income tax – to tax the rich.)

To be sure, there is a lively debate about the significance of technology in the rising income-wealth gap and without question there are other factors. The decline of organized labor was once an equalizing force in our economy as it enabled millions of people with marginal education and skills to move into the middle class. But hostility to unions has fed a steady decline of union ranks. Many states have adopted "right-to-work" laws that outlaw union shops in which every employee of a unionized company is required to pay union dues.

In the early half of the 20th century, labor unions were a major force in the nation's economic and political life. In the early 1950s, about a third of all workers were in unions. But then the long decline ensued. Today the overall union membership is about 10 percent, and only about 5 percent of private sector workers are organized. Some 40 percent of government workers are organized which reflects largely the lack of political will to resist unionization.

The decline of labor has had a profound impact on the income gap. A half century ago the major auto companies accounted for about 7 percent of the work force. A typical employee of General Motors, the nation's largest employer at one time, backed by the United Auto Workers made $35 an hour. It is actually much more than that today, but now the largest employer is Wal-Mart where an average worker earns about $9 an hour.

The reality on the ground today is that one out of every three working Americans is in a part-time job, sometimes more than one. Some of these people of course are doing quite well working as consultants, free-lance writers, and independent contractors repairing homes or performing landscaping services, that sort of thing. But many more are eking out a marginal existence living from paycheck to paycheck, and digital technology is the main reason.

"It just seems so obvious to me that technology is accelerating the rich-poor gap," said Steve Jurvetson, a venture capitalist at DFJ Venture in Menlo Park, California. In discussions with other in the high-tech community, he says this impact of technology has long been "the elephant in the room, stomping around, banging off walls."

"My reading of the data is that technology is the main driver of recent increases in inequality," said Erik Brynjolfsson, a professor of management at the Massachusetts Institute of Technology's Sloan School. "It's the biggest factor." Brynjolfsson (please don't ask me to pronounced that name) contends that innovation is rapidly accelerating as trends in computing and networking advance at an exponential rate. But "while the pie is increasing," he said, "the pieces of that pie are not being evenly distributed."

According to Colin Gordon writing in *Dissent,* technology and inequality are linked by two apparent mechanisms. The first is the capital bias of recent technological change which leads to rapid automation. This tends to displace labor and deliver greater returns in productivity. The second is the skills bias of recent technological change which has outpaced educational achievement. The demand for skills has outpaced its supply. This had tended to hike the wages of workers with advanced skills (or at least the right skills) while leaving the lesser educated and less well trained behind.

Stated another way, advancing digital technology is hollowing out the labor market as robots and computers are displacing middle-wage jobs and reserving job growth for tech-savvy jobs at the top and low-wage service occupations at the bottom which are more resistant to advanced technology.

"All of this, unfortunately, brings us to something of a political dead end," wrote Gordon. "Technological innovation or change is not something you can readily control, so there seems little recourse but to occasionally lament the quality of American education or the unfortunate prevalence (in the recent recovery, or in the long term) of McJobs. Those who design or program the next Roomba will get rich; those displaced by it will not."

In fairness, for all of his astute analysis of the impact of technology on the wage gap, Gordon remains unconvinced that it is the main culprit. "If the gap between winners and losers was really about technological innovation or skills, we would expect to see a close correspondence between wages and educational achievement," he wrote. "Yet while those with a college education or better pulled away from the pack in the 1980s and 1990s, that advantage has slowed dramatically. Wage growth has been flat since the late 1990s for all educational cohorts. And workers with a college degree have lost nearly as much ground since 2007 as everyone else."

Gordon cites the work of Thomas Piketty and Emmanuel Saez demonstrating that wealthy nations with similar histories to ours of technological change show divergent patterns of inequality. They contend it is institutional and policy differences driving inequality more than technology. "In the United States, those differences – or those lost protections – are now pretty familiar," Gordon wrote, "a long decline in union density and bargaining power, the retreat of basic labor standards and their enforcement, a fiscal policy that accommodates spells of high unemployment out of an irrational fear of inflation, and the evaporation of fiscal or regulatory restraint on incomes at the very top. The robots and computers had little to do with it."

I agree with Gordon that much of the income inequity is born of unwise public policies and I would add one more – an educational establishment out of synch with the real world workplace. His observation of college graduates losing ground is right on point. Thousands of them are out there tending bar or driving taxis because their expensive college educations have not prepared them for viable opportunities in the world we live in.

But then the main reason for that is that institutions of higher learning are not staying abreast of the technological revolution. They use high tech to great effect in the classrooms and the students pack laptops everywhere they go, but the content of what they teach has changed scarcely over the generations. Their graduates are for the most part inordinately unprepared for life in the real world.

In the end, the dispute about the contributing impact of digital technology on the income gap comes down to the old Chico Marx line, "Who you gonna believe, me or your eyes?" We can plainly see the impact of technology on the workplace and worker wages. Employers spend lavishly on advanced technologies because they promise higher quality products and lower production costs. "Automation and digital technologies have reduced the need for many production, sales, administrative and clerical jobs," wrote David Rotman in *www.technologyreview.com.*, "while demand has increased for low-pay jobs that can't be automated, such as those in cleaning services and restaurants." The result is a "barbell-shaped" job market with strong demand at the high and low ends and a hollowing out of the middle. And though there are ample low end jobs, there are too many applicants which tends to drive down wages in those categories as well.

Actually, assigning responsibility for the growing income disparity is not an either-or choice. "Emphasizing technology does not deny that changes in labor market institutions have been important," wrote Daron Acemoglu in the *National Bureau of Economic Research Reporter.* "The erosion of the real value of the minimum wage and the declining of unions undoubtedly have been important for changes in U.S. inequality, especially at the bottom of the wage distribution. In addition, the late 1980s and the 1990s have seen an explosion of CEO pay, which is difficult to explain with changes in technology alone, and which suggests that there may have been concurrent changes in social norms pertaining to inequality and fairness."

It is worth noting that some of the most glaring contrasts in wealth are abundantly visible in the birthplace of the digital revolution – Silicon Velley. Median income in Silicon Valley reached $94,000 in 2013, far above the national median of around $53,000. "Yet an estimated 31 percent of jobs pay $16 per hour or less, below what is needed to support a family in an area with notoriously expensive housing," wrote Rotman. And yet the poverty rate in Santa Clara County, the heart of Silicon Valley, is around 19 percent according to calculations that factor in the high cost of living.

"You have people begging in the street on University Avenue (Palo Alto's main street)," says Vivek Wadhwa, a fellow at Stanford University's Rock Center for Corporate Governance and at Singularity University, an education corporation in Moffett Field with ties to the elites in Silicon Valley. "It's like what you see in India," said Wadhwa who was born in Delhi. "Silicon Valley is a look at the future we are creating, and it's really disturbing."

The clash between wealth and poverty in Silicon Valley is striking. The wealth generated in Silicon Valley is "as prodigious as it has ever been," said Russell Hancock, president of Joint Venture Silicon Valley, a nonprofit group that promotes regional development. "But when we used to have booms in the tech sector it would lift all boats. That's not how it works anymore. And suddenly we're seeing a backlash and people are upset."

Yes, and I am one of them. This great digital revolution was supposed to remake our world and usher in another round of robust economic growth built on growing opportunity and personal wealth. The promise is not being fulfilled, not by a long shot.

# CHAPTER ELEVEN

# Undermining Democracy

*The people have spoken, the bastards.*
*—Dick Tuck*

When Robert E. Lee surrendered the Confederate Army to the Union forces under General Hiram U. Grant in 1865, ending the Civil War, there were many diehards among his troops who were determined not to give up the fight. Certainly, they still had tens of thousands of armed soldiers ready to continue. Their idea was to take to the hills and launch a guerilla war. There were more than a few rebel leaders well versed in that kind of warfare – such as Nathan Bedford Forrest and John Mosby, alias "The Gray Ghost." They could have kept it up indefinitely.

But General Lee, to his everlasting credit, would have none of it. He spoke of the devastating impact such a prolonged, endless conflict would have on our country after what was already the bloodiest four years of our history. He knew the toll it would take on the nation, especially his beloved southern states that were already in ruins. He thanked his loyal officers for their courage and sacrifice and urged them to accept the results. He exhorted them to return home to their families and farms and to be good citizens.

I have always cherished that account of Lee's moral courage and vision that spared our country many years of needless suffering and set the stage for eventual reconciliation among the states. At West Point we have a lovely pathway along the Hudson River called Reconciliation Walk, sometimes Flirtation Walk, where cadets can stroll with their romantic interests. Of course, West Point produced the leaders who fought on both sides of the Civil War. In the years after the war, the academy struggled to incorporate new generations of southerners into its ranks, leaving the past behind.

There is no direct connection between Reconciliation Walk and the Civil War, but to me and many of my fellow cadets, it symbolized that period of our history, and our mission to heal the nation's wounds and fight together in its defense in the wars that lay in our future.

Today we need another good dose of reconciliation. We live in an era of political polarization in which our citizens are increasingly isolated from each other in basically two warring camps, unable to recognize and respect the other side's points of view and resistant to any suggestion of compromise. It seems that virtually every issue comes down to "them versus us" in one way or another. Polarization is a conspicuous aspect of the American political scene that is abundantly obvious to all – including the politicians who must somehow find a way to bridge the gap to win election. President George W. Bush famously described himself as "a uniter, not a divider." Predictably a public opinion poll a few years later found that 49 percent of Americans thought he was a "uniter" and 49 percent thought him to be a "divider." His successor President Obama likewise strove mightily to forge a bipartisan coalition but fell on his face.

Not since the pre-Civil War era when our nation was divided between slave and free,

hurtling toward a brutal confrontation, have we seen such a rigid polarization of opinion. But at least back then there was a clear issue dividing our ancestors. Today, the schism seems built upon an array of issues that more or less just happen along a political fault line that divides people over gun control, taxes, race relations, abortion, gay marriage, gun control, foreign trade, immigration, and the general economic malaise. One would think there would be a mixture of opinions among the different partisans – say gay rights activists opposed to tax increases or second amendment champions who support liberal immigration – but it doesn't seem to work that way. With a few exceptions here and there, most people are singing from the same hymnbook depending on whether their sympathies lie with the right or the left. It makes me wonder if they really feel that strongly about specific issues or are simply going along with "their" crowd because they want that sense of belonging.

Whatever the reason, the divide is real. "Republicans and Democrats are more divided along ideological lines – and partisan antipathy is deeper and more extensive – than at any point in the last two decades," said the Pew Research Center. "These trends manifest themselves in myriad ways, both in politics and in everyday life."

Pew said the divisions were greatest among those most

engaged in the political process. The overall share of Americans who express consistently conservative or consistently liberal opinions has doubled over the past two decades from 10 percent to 21 percent. "And ideological thinking is now more closely aligned with partisanship than in the past," Pew said. "As a result, ideological overlap between the two parties has diminished. Today, 92 percent of Republicans are to the right of the median Democrat, and 94 percent of Democrats are to the left of the median Republican."

Partisan animosity has increased substantially over the same period, Pew said. In each party the share with a highly negative view of the opposing party has more than doubled since 1994. Most of these intense partisans believe the opposing party's policies "are so misguided that they threaten the nation's well-being."

Pew said the ideological divide is not necessarily shared by all or even most Americans. The majority do not hold uniformly conservative or liberal views, or see the other party as a threat to the nation. Most people would like to see their representatives make compromises to resolve contentious disputes. "Yet many of those in the center remain on the edges of the political playing field, relatively distant and disengaged, while the most ideologically oriented and politically rancorous Americans make their voices heard through greater participation in every stage of the political process."

In my younger days, it was generally considered positive that people felt strongly about their beliefs and worked to advance them through the political process. But today that has somehow evolved into a destructive force that leaves our government unable to make the basic compromises demanded by democracy.

One would think also that such bitter political partisanship would lead to dominance of one side or the other as it generally has in years past. During the Great Depression, Franklin D. Roosevelt seized the White House and led the Democrats to dominance of Congress that lasted almost 20 years. In 1980, Ronald Reagan led the Republicans to dominance that lasted more than a decade. But the polarization today is fragmented in strange and unfamiliar ways that give the White House to Democrats and the Congress to Republicans – and never the twain shall meet. There is something new and disturbing going on.

Of course, in the volatile campaign season of 2016 it seems a stretch to blame the partisan divide on opposing views about specific issues, at least on the Republican side, because the leading candidate (as of this writing), Donald Trump, has no discernible ideology or philosophy, or at least not one than can be associated with Republicans or Democrats, conservatives or liberals. His message is in essence not a message at all but rather a rant about the world as it is that strikes a resonant chord among his followers. Many people out there feel they have been shunted aside, their jobs have been exported and their core values are under attack. Exposure of his misstatements, mistakes and bigotry does not discourage Trump's followers because it does not address their angst. Rather, it is because Trump is so estranged from the "establishment" that has somehow failed them that they flock to his banner.

There is a steady stream of commentators and academics burning up their computers trying to define the causes of this unusual partisanship, but their opinions span a large realm of discussion and provide no more consensus than the political parties can achieve. Much of it is surely due to simple arrogance – an assumption that I am right and you are wrong – but that seems to be a timeless human failing certainly not unique to our time.

I believe the answer to this riddle lies not in ideology, conflicting values or even politics itself but rather the digital technology that is reshaping virtually every aspect of the way we live, work and get along with each other. Why would we think it is not also having an impact on our political system?

## Influencing Opinion

Not so long ago, our political power structure was defined by public opinion which in turn depended on the dissemination of information through the major news media. There were three major TV networks – CBS, NBC and ABC. There was also a smattering of public channels but the three big networks ruled the roost. Those who depended on TV for the news, and most everyone did at least to some extent, went to one of those three conduits. They basically covered the same news from the same mildly liberal enlightened point of view.

But the news was generally balanced and everyone who watched it inevitably came into contact with unfamiliar information and viewpoints that challenged their own. Sometimes it made you angry. I know it often made me angry. But merely being exposed to opposing points of view made us think and often reconsider our views. Perhaps you did not change your mind or alter your core beliefs, but at least you were aware that other people out there saw the world through a different lens, or that at least some of the facts did not support your position. Often this experience led to personal exchanges between people at work or in school that further challenged us to defend our views and perhaps alter them in the face of articulate opposition. On some occasions, this sort of thing led to fist fights, but even then there was actual thought in action. To deal with opposing points of view is simply basic to intellectual growth and acquisition of understanding. We need it to grow.

But the new digital technology for better or worse has broken the chokehold those three major networks once had on public discourse. They are still there, still providing mostly objective journalism, but their audiences are much smaller than in days of yore. We all today enjoy many more options regarding where we acquire our information and there is an almost irresistible temptation to get it from people who reflect our personal prejudices. Thus if you are of a conservative bent you are probably watching Fox News and if you lean more to the left you are watching MSNBC. Of course, those are the big ones. There are other options. You can get your news through a religious filter if you prefer or news from China or Europe. The options are virtually endless.

And of course, there is the ever present business news going 24/7 with constant updates on the stock markets and discussion of likely investments. In 1982, there were no such things. Cable TV was just being born. The U.S. Chamber of Commerce came up with the first solid business reporting in a two hour broadcast aired in the morning five days a week on the new sports network ESPN which in the early days was having trouble filling up its air time. But to hire staff for this new adventure the U.S.

Chamber, a very conservative organization, had to hire professionals away from established TV outlets, all of whom came with that ubiquitous enlightened liberal attitude toward the world. Thus, my good friend Jeff Joseph, who was in those days a U.S. Chamber Vice President, presented daily commentaries from a more conservative, pro-business point of view. He was like the little Dutch boy with his finger in the dike, or so it seemed to him. But in less than 10 years that fledgling program was overrun by the all the business news all the time networks. All made possible by digital technology.

The same transition held true for the other timeless medium – radio. Only a few years ago you could turn on the radio driving to work or perhaps doing chores around the house and get a balanced feed of national news, local news, weather, sports, community events and lighthearted banter. The people who were good at radio thrived in an atmosphere that encouraged and rewarded responsible creativity. Radio was a great contributor to the quality of life for almost a century. Today one is hard pressed to find any radio outlet offering anything other than infomercials and nonstop political harangues, usually from the political right.

President Obama captured the flavor of what we are living with in a YouTube interview with Destin Sandlin, creator of a popular video series on science. "Some people are just watching Fox News; some people are just reading *The New York Times*," he said. "They almost occupy two different realities in terms of how they see the world." He lamented the absence of a "common baseline of facts" underpinning the political debate and accused the Republicans of peddling – through their own information channels – an "alternate reality" on important issues such as climate change and the economy.

I would tend to agree with President Obama on that, but would add that I see scarce difference between the Republicans and Democrats on this tendency to choose one's facts selectively and to dismiss out of hand opposing views. Partisanship is not a partisan affliction, it is universal and it is made possible, indeed abetted and encouraged, by digital technology.

President Obama also asserted, during a town-hall style event in Buenos Aires, of all places, that smartphones were "isolating" people. He said that with the new smartphone technology, young people "just surf the surface of information as opposed to analysis and understanding and study." He said the public too often takes in "just the Twitter line without trying to figure out, okay, is this true or not? What are the facts?" Obama was right about that.

I hasten to mention that advanced digital technology has enabled partisanship by giving political activists at the state level a powerful device for redesigning Congressional districts to achieve the maximum advantage for the party that controls the state legislature. The issue is an old practice known as Gerrymandering that has achieved unprecedented virulence. Here in the nation's capital we are sandwiched in between Virginia and Maryland, two of the worst offenders. Maryland's voters are split about 60-40 in favor of the Democrats over the Republicans, but you wouldn't know that from the state's Congressional delegation which consists of one Republican among seven Democrats. The situation in Virginia is the reverse of Maryland with a slight edge for Republicans among the electorate but only three of eleven seats held by Democrats against eight Republicans.

All across the country it is the same dismal story – computer designed congressional districts that empower one party at the expense of another, lumping most opposition voters into concentrated districts. A perverse result is that most incumbents, in order to get reelected, cater to the extremes of their own party. Precious few seek out a middle road in an effort to reach out to both parties. Ergo, when they get to Washington, they embrace extreme positions demanded by their most partisan constituents.

Democracy has always been a fragile reed, dependent on the ability of credible people to influence public opinion with reasoned analysis of the challenges facing society. Human beings tend to be opinionated and stubborn in the best of times. When their attention spans are abbreviated by cell phones, their ability to think is compromised. They are instead looking for confirmation of their biases. "People are going to resist the suggestion that their candidate is wrong," said Bill Adair, a journalist who launched the Pulitzer Prize-winning PolitiFact website at the Tampa Bay times in 2007. "But what is different now is there are more partisan outlets that reinforce the partisan feelings."

Yes there are, and they are made possible by the revolutionary new digital technologies which are at least indirectly responsible for the Donald Trump phenomenon. Trump is a demagogue playing to people's credulity like P.T. Barnum and their fears like Huey Long. A typical young American today would not even recognize those names. They cannot learn from the past because they know little about it. It just doesn't come through Facebook or Twitter. There are legions of professional journalists out there doing their utmost to convey the truth, but their reasoned analysis cannot compete amid the din of partisan propaganda beamed forth in the cacophony of media outlets to people with abbreviated attention spans. The absurd 140- character limit of a Tweet is perversely receptive to the incoherent ramblings of a demagogue, and we have many in our midst.

On a more fundamental level, we are losing sight of our core values – faith, family and friends – that are the glue that holds us together. The digital technologies have a perverse tendency to drive us apart, enabling us to live as individuals instead as members of our communities, churches and neighborhoods. The proliferation of deceit, meanness and smut on the Internet has a corrupting effect that debases our culture and undermines our democratic values and processes. It is a graver threat to our security than the terrorists could ever pose.

"When a population becomes distracted by trivia, when cultural life is redefined as a perpetual round of entertainments, when serious public conversation becomes a form of baby-talk, when, in short, a people become an audience, and their public business a vaudeville act," wrote Postman, "then a nation finds itself at risk; culture death is a clear possibility."

# CHAPTER TWELVE

# Digital Technology To The Rescue

*Success covers a multitude of blunders.*
*—George Bernard Shaw*

My readers – at least those who have persevered thus far – deserve a break from this apparently endless screed about the failed promises of digital technology, and thus I interrupt the flow to acknowledge that yes, there are some occasions and instances in which mankind is making effective use of this great leap forward. We have without question found creative ways to squander our genius, but now and then and here and there a few of us have stepped forward, put the nonsense and greed aside, and used the new technology to positive effect.

An obvious case in point is the military where I served for 35 years. As an officer in the Signal Corps, and eventually the senior officer in the Signal Corps, I played a role in helping the U.S. Army stay abreast of rapidly changing communications technology from 1953 to 1987, and in later years made additional contributions as a consultant and commentator. In the Korean War, I had carrier pigeons as a backup in case our primary means of communication, basically wires laid along the ground, were interrupted. I tested them on one occasion and most of those little critters actually got through. The carrier pigeon program was eventually shut down in 1957 – the U.S. Army is slow to abandon traditions. In the 1980s, I helped the Army and other services make the transition from analog to digital and begin adopting the new technologies into our defense systems. A major consideration then and now was the fail-safe systems on our nuclear arsenal which thus far at least have enabled us to avoid disaster.

## Operation Desert Storm

More conspicuous results of our work were on vivid display in Operation Desert Storm when Iraq under Saddam Hussein invaded Kuwait in 1990. This was one of those rare events in history when responsible leaders exercised good judgment. Secretary of Defense Colin Powell did not react precipitously. Under his leadership over a period of months we assembled a vast military machine including 540,000 U.S. troops, including our advanced armor and aircraft and naval vessels, and assembled a broad coalition to help us fight the good fight. We won the war handily. ( I cannot resist mentioning that President George H.R. Bush had the good sense not to move into Baghdad and take over the country because he realized our great victory would quickly devolve into a quagmire, but that's another story.)

A year later, in July 1991, the Senate Armed Services Committee conducted a thorough review of the war and identified several factors that contributed to our overwhelming victory: the high quality of U.S. commanders and their personnel; tough, realistic training; streamlined joint command relationships; and – last but certainly not least -- sophisticated technology.

The latter item was a source of tremendous pride to me. That sophisticated technology included some of the best communications gear ever provided to U.S. military personnel, or any military personnel. The communication network that supported Operation Desert Storm was the largest joint theater system ever established. We pulled it together quickly and kept it operational 98 percent of the time, which is unheard of where computer technology is concerned. At its peak, the system supported 700,000 phone calls and 152,000 messages per day. More than 30,000 radio frequencies provided connectivity to ensure minimum interference. It was an astounding achievement.

Satellites played a key role enabling U.S. land-based forces to go from almost nothing to an extensive tactical network in the region. The Persian Gulf area was a black hole almost devoid of U.S. global communications other than with the Navy. The Defense Communications System essentially stopped at Turkey in the west and the Philippines in the east. Commercial communications in the region were limited and military systems were only beginning to come online.

But by the end of October 1990, the basic command and control structure was in place. Additional communications capacity and European gateways were added to support the larger force structure. Increased connectivity into Europe was needed to tie the VII Corps to its sustaining base in Germany. Desert Storm was the first major sustained military operation in the microprocessor era. Reliance on computers grew as the operation continued. People who used computers in garrison took them along into the field. Thousands of computer terminals were used, most of them off-the-shelf PCs. They held up well in the desert with sensible maintenance.

All of the services – Army, Marine Corps, Navy, Air Force – operated aircraft and fielded air defense systems in the Persian Gulf region during this conflict. A key challenge was to assemble a system enabling the aircraft of different services and different nations to fly to and from their target assignments without conflict with each other. The scheduling of air tasking was one problem and the real time control of the aircraft was another. AWACs, E-2C, Aegis, Hawk, Patriot and ground-based surveillance radars tracked and de-conflicted the numerous air contacts providing identification of friend or foe, guarding against friendly fire.

In this way, several generations of technologies were made interoperable for airspace control. The result was a complex integrated system that supported up to 3000 combined air sorties each day, and controlled more than 25 air defense sites and six carrier battle group air defense platforms.

In the end, we won handily because our commanders had a picture of the battlefield and the enemy did not – all made possible through our advanced digital communications systems. The advanced aircraft and tanks that did the actual fighting get most of the credit, and they deserve it, but the communications, satellites, sensors and computer systems that facilitated the integration and use of such sophisticated weaponry with such accuracy and impact, and without major accidents, deserves much of the credit.

Technology was adapted quickly to meet operational needs. The technology and the people were flexible enough to react to changing requirements of the situation.

The folks at home – including me – got to watch some of this on television as we were treated to scenes of row upon row of military personnel in tents glued to their laptops as they acted their parts in a highly complex system bringing all of these disparate technologies to bear in one great synchronized battle formation. But even an old warrior like me must lament that it is disappointing that we can master this technology so effectively in war, but not in peace.

## Modern Manufacturing

"The world power that loses its manufacturing base," said Akio Morita, founder of Sony, "will cease to be a world power." Morita was right. Manufacturing, more than any other sector of the economy, is where the rubber meets the road. It is where raw materials are taken from the earth and forged into the wonderful labor saving products we depend upon in daily life. It is modern day alchemy – transmuting base metals into things more valuable than gold.

The post-industrial myth that we lost all of our manufacturing strength is due to the huge loss of manufacturing jobs in recent years. While millions of jobs were lost, they were primarily low skill jobs in the old manufacturing model. The perception that we have lost most of our manufacturing jobs is way off base. By the most conservative count conducted by the Bureau of Labor Statistics, we still have more than 12 million manufacturing jobs. That is a lot of jobs by anyone's definition and a variety of data confirm that manufacturing jobs pay significantly more than other jobs.

The reality is that manufacturing represents a much larger headcount than the BLS would have us believe. Those 12 million manufacturing jobs support more than half as many more jobs in other sectors, which would put the total job count directly attributable to manufacturing well north of 18 million.

In other words, manufacturing is alive and well in the U.S. and the main reason is advanced technology which is the major factor defining the innovation in the new manufacturing. Two-thirds of R&D funding is in manufacturing and the majority of patents – about 90 percent -- are acquired in manufacturing.

There is no great mystery why manufacturers are so focused on technology and innovation. Manufacturers must compete on the global stage. They are under tremendous pressure to constantly improve quality and reduce costs, and also to bring new products and processes to market. The factory floor is the best place to find out if new ideas actually work as intended and work out the kinks.

A major key to growing competitiveness of U.S. manufacturing is increasing reliance on advanced robotics, which is at last fulfilling the imagination of science fiction writers. Some factories and distribution centers today are almost empty of people as robots scurry to and fro performing a variety of functions in closely timed synchronization.

We are currently making and selling about 200,000 robots a year. Robots of course make a tremendous contribution to the quality revolution because they don't make mistakes. You are seeing more robotics, in the production of auto, healthcare products and services, hotel service and many other areas.

Additive manufacturing –sometimes called 3D printing– represents another sea change in the way we make things. The old way was subtractive manufacturing – you take a block of metal, wood or something else and chip away with stamping machines and lathes until you have the desired shape you want for whatever purpose. Additive manufacturing turns this process on its head. It creates three-dimensional products based on computer files by sequentially depositing thin layers of liquid or powdered metals, polymers or other materials on a substrate.

Additive manufacturing can encompass metals, polymers, and electronics, and can apply to a range of structural and functional materials and to a range of components for defense and energy applications. Parts can be fabricated as soon as the three-dimensional digital description of the part needed is created. This is creating a new market for on-demand, mass customization manufacturing. Novel components and structures can be produced from additive manufacturing processes that cannot be cost effectively produced from conventional manufacturing processes such as casting, molding and forging.

To better understand the speed of digital; technology change we should go back and visit "The Kodak Moment." In 1996, Kodak had 140,000 employees and a $20 billion market cap, and controlled 90 percent of the film and camera market. But by missing the importance and speed of digital technology, Kodak began to lose money in the latter part of the 90s, stopped turning a profit by 2007, and filed for bankruptcy in 2012.

Manufacturing companies have quickly learned the power of digital technology, and Moore's Law about the speed of change. The use of digital technology became the leading driver of the manufacturing comeback. The digital revolution has now been with us for a while in creating everything from sophisticated consumer products to highly productive machines to complex supply chains.

The major world corporations are today interconnected like never before and that enables senior corporate managers to monitor company activities daily – hourly, by the minute – on the factory floor, even if that factory floor is in a foreign land. It enables them to monitor, respond to and anticipate consumer needs and desires, wherever that consumer is a customer in a store or a vendor buying robots.

We are embarked upon a new Internet era that aims to integrate the full range of corporate activity. Some call it the Internet of Things – or the IoT. Cisco CEO John Chambers called it the "fourth wave of the Internet." Simply put it is "the marriage of minds and machines."

Whatever you call it, it means corporate management is finding creative new ways to employ digital technology to improve quality, productivity, and the bottom line. The IoT will provide senior management with a steady flow of information that will take much of the guess work out of running an organization. It will also minimize the threat of surprises – such as power outages and flight delays.

But other more significant factors are at work. Because of advances in gadget wizardry and more exacting consumer demand, many companies are finding it more convenient to make their products closer to home. The cycle life for many major consumer products such as refrigerators and ovens has been shortened from seven years or so to two or three years as more innovations improve the product and make it more attractive to consumers. If you are going to be upgrading your product every two or three years, you need to be closer to production, and you need to be able to work directly with the people on the production lines – which can be a problem if they are thousands of miles away and speak a different language.

Along with disruptive technologies come disruptive skill requirements and shortages. Manufacturers around the country are loudly lamenting the difficulty they have finding applicants qualified to work in modern manufacturing. 80 percent of manufacturers report mild to severe skill shortages, with estimates of the magnitude ranging from 1 million to 2 million jobs depending upon the definition and timeframe of the analysis.

The causes of the manufacturing skill gap ranges from the negative perception of manufacturing, to the retirement of the baby boomers, to the lack of basic skills among many new workers entering the labor force, to the need for more advanced skills associated with STEM: science, technology, engineering and mathematics. Today's manufacturing company requires a high performance workforce to achieve both basic and advanced manufacturing production and distribution.

It is a truism that manufacturing is evolving into something new and that it will never again be the incubator of millions of middle class jobs for people with limited skills. But it will be an incubator for millions of more sophisticated jobs that bright young people will be increasingly attracted to as they come to appreciate the high tech environment of modern manufacturing.

## The Internet of Things

The Internet of Things (IoT), machines talking to each other, bypassing the unreliable human element, is generally associated with manufacturing because that's where most of the advanced machines are and where there is intense pressure to raise quality and productivity. It is a lot more than simply machines talking to each other. A machine is merely an instrument. To fully realize the potential of IoT, sensors must come into play. The sensors measure and evaluate for the machines. Together they comprise a whole new enterprise.

That's where the cloud come sin. Cloud-based applications are key. The IoT cannot function without cloud-based applications to interpret and transmit the data coming from all the sensors. The big impact comes when IoT data inform decisions, enabling managers to achieve higher levels of efficiency and productivity. In factories, sensors will make processes more efficient providing a constant flow of accurate information to managers. Sensors that tell management when equipment is wearing down and nearing the end of its useful life can reduce maintenance costs and down time. They can automatically update inventories. The potential applications are limitless.

Outside the factory gates, IoT is being used in health and fitness devices – telling you how many steps you have taken or calories burned – and making possible the new concept of "smart homes" that manage themselves, turning down the lights and heat when no one is there, notifying police or fire departments when something goes wrong.

In sum, the IoT has tremendous potential to boost productivity and enhance the quality and length of human life. The potential economic impact is in the trillions of dollars and the reality on the ground is moving rapidly. It will be really hard for us to screw this up.

## Estonia

In discussing success stories about the effective use of digital communications technology, I would be remiss if I did not mention a sovereign nation that has seized the new technology in a wholehearted way – Estonia. Estonia is not a nation that generates a lot of news. It is one of three small countries perched on the edge of the Baltic Sea in Northern Europe. The other two are Lithuania and Latvia. You might assume that these three little nations are somehow related in language and culture. But they have distinct languages and cultures that somehow seem to have no direct relationship with each other. When people from Estonia wish to converse with people from Lithuania or Latvia, they speak either English or Russian.

The three little Baltic States are former Soviet Socialist Republics that escaped the Soviet embrace a little over 20 years ago. They are all immensely relieved to be free of the Soviet yoke, but understandably uneasy about their security, given the incessant rumblings out of Moscow. In years past, the Soviet Union, and before that Russia, frequently violated Estonia's sovereignty. The aggressive policies of Vladimir Putin keep their anxieties alive.

But they are determined to preserve their freedom and

independence. Estonia has only about 1.3 million citizens – about half the population of Chicago. But it is an active member of the North Atlantic Treaty Organization (NATO) and is in fact one of only five NATO countries (out of 28) to spend the NATO mandated 2 percent of its GDP on defense. When other NATO countries timidly sent their troops to the safer parts of Afghanistan – acknowledging that safety is a relative term in Afghanistan – Estonia sent its people to Helmand province, one of the deadliest regions. Estonia wants other NATO members to know that it is a full-fledged member, not a free rider.

Estonia is frequently touted as the world's premier example of a country with a strong technological infrastructure and commitment to Internet freedom. In Estonia, voting, signing documents and filling out tax returns is done online through X-Road, an online tool that coordinates multiple online data repositories and document registries. X-Road provides all Estonians – ordinary citizens, enterprises and government officials – with unparalleled access to the data they need to do business – quickly obtaining licenses, permits and other documents that can take weeks or months in other countries.

X-Road is built with scalability in mind so that the growing number of services and repositories can easily be attached to the system. Although this digital backbone alone is rather impressive, it is just one of many products in a country ahead of the digital curve. Instead of being held back by its past and falling victim to ailments that plague many post-communist countries, such as corruption, bloated government and an obsolete education infrastructure, Estonia has decided to start with a clean slate and invest in its future. To transform its society in a community of tech-savvy individuals, children as young as 7 are taught the principles and basics of coding. By comparison, only one in four public schools in the U.S. teaches computer programming.

All of this creative infrastructure serves to promote the entrepreneurial spirit. According to The Economist, in 2013 Estonia held the world record for the number of new company startups per person. Many of these startups are successful including recognizable names such as Skype, Transferwise, Pipedrive, Cloutex, Click & Grow, GrabCAD, Erply, Fortuno, Lingvist and many others. Skype was actually invented by a Swede but he came to Estonia where it was developed into a working system for advanced communications. It was purchased by eBay for $2.6 billion in 2009.

Estonia even goes out of its way to invite foreign entrepreneurs to set up shop there without actually moving there. Through Estonia's e-residency service, a transnational digital identity available to anyone, you can not only establish a company in Estonia via the Internet, but they can also have access to other online services that have been available to Estonians for over a decade. This includes e-banking and remote money transfers, declaring Estonian taxes online, digitally and signing and verifying contracts and documents. E-residents are issues a smart ID card, a legal equivalent to handwritten signatures and face-to-face identification in Estonia and worldwide. The cards themselves are protected by 2048-bit encryption, and the signature/ID functionality is provided by two security certificates stored on the card's microchip.

The Estonian health care system is online so doctors can review patient files or track national health trends more efficiently. Imagine that. Most Estonians choose to leave their medical records accessible to medical practitioners, but each patient can see the names of every visitor to his or her file, and can block unwanted visitors, or lock the information. Estonians can also pay their taxes and even vote online. They are gradually expanding their Internet to daily systems and devices, trying personal networks to cars, televisions and even refrigerators.

Estonia is acutely aware of the threats of cyberattacks having been assaulted by hackers in 2007, most likely from Russia. The country responded quickly enlisting all citizens to help the nation develop defense tactics. Everyone in Estonia has to participate in their interactive system. They see computer technology as the wave of the future and realize to survive among the large nations, they need to be ahead of the technology curve. They even have sixth graders learning to write code.

Estonian President Toomas Hendrik Ilves, during a recent visit to George Washington University in Washington, D.C., said cyber capabilities are vast in a world that has become dependent on technology. The Internet can manage everything from a country's power plants to milk deliveries, said Ilves. He noted that individuals are at risk of revealing sensitive personal information through big data. "We've gotten to the point where you don't need to physically attack a country to debilitate it – we've seen this ourselves," he said.

Estonia has addressed security threats by creating protected online identities and issuing digital signatures that have allowed for well-guarded e-services, like e-voting and electronic tax filing. The country has also taken measures to secure individuals' data. "There are no rules," said Ilves, who is himself a cyber security expert, adding that new solutions must be established between society and government, and data security must be guaranteed.

Estonia's workforce in both the public and private sectors has come together to build digital security into their system. It is reflection of Estonia's leadership that the NATO Cooperative Cyber Defence Centre of Excellence is housed in Estonia's capital, Tallinn, and cyber strategy is foremost on the Estonian parliament's agenda. Estonia's cyber-infrastructure, known as "e-Estonia," is second to none.

The Maryland Estonia Exchange Council (MEEC) is actively promoting more extensive ties between Estonia and the Free State. Originally launched by the Maryland National Guard, MEEC has helped foster several sister-city relationships between cities in Estonia and Maryland. "We seek out cities in both countries and encourage them to establish sister relationships," said Erik Pusker, a member of the MEEC board. Pusker said the connection was being extended to universities and business organizations.

Pusker noted that when Estonia broke free of the old Soviet Union it was saddled with an inefficient smokestack industries that the Estonians realized would be shut down anyway. They have instead embraced high tech industries. For example, Estonia's archaic telephone system that it inherited from the Soviet era would have required massive investment to modernize. "They just skipped that phase and went straight to reliance on mobile phones," he said. "That enabled them to leapfrog many things."

In sum, tiny Estonia is showing us – and the world – how thoughtful citizens can embrace modern digital communications technology and make it work for them at all levels, from medical care to national defense. A country must be willing to adapt and change the infrastructure of both the government and the economy as needed, and to continually improve them. An intense focus must be placed on the educational system, and entrepreneurship must be encouraged.

Of course, Estonia is a small country and given its peculiar situation, was able to basically begin from scratch to build its digital framework. But in doing so it is offering an example of how the digital technology can be used to modernize a nation both culturally and economically. Estonia is showing us the way forward through the digital revolution, showing in practice how its potential can be harnessed to benefit the people of a forward thinking nation.

## EPILOGUE

*Tis grace hath brought me safe thus far,*
*and grace will lead me home.*
*—John Newton*

Over my 65 years working through, and sometimes leading, the communications revolution I have had a sobering
reminder of the frailty of our species, and our difficulty
adapting to change, especially upheavals such as that
wrought by digital technology. As our experience of the
late 19th century demonstrated, it is not enough to discover
revolutionary new sources of power such as that provided
by steam, electricity and the internal combustion engine. It
takes a mighty long time to fully harness such power, to
make it work for us instead of against us, to discourage its
misuse and abuse, and to accommodate its sometimes brutal impact on our economy and society. We are still in the
early stages of the digital revolution, still working out the
kinks and groping for ways to make this new power a positive force in human affairs. And of course there are many
working in the shadows to use the new power for disreputable purposes.

The quest to harness the new technology does not occur in a vacuum. As of this writing, we are embarked upon
another one of our periodic uproars over ethnicity – this

time focused on admission of people from Islamic countries and the presence of some 11 million illegal immi-
grants in our midst, most of them of Hispanic origin. This
is not new. In the 19th century, many businesses posted
signs saying "No Irish Need Apply." Many employers and
social organizations embraced policies discriminating
against Jews. And of course not until the 1950s did we
finally begin to come to grips with our shameful history of
abuse of African-Americans and despite the election of an
African-American to the Presidency, we still have a long
way to go to reach any sort of nirvana of racial harmony.

Now here we go again. We do not need walls between
the United States and Mexico or any other country. As a
soldier, I learned early that fixed fortifications are monu-
ments to the stupidity of man. Before I began to build a
wall, I would want to know exactly what it was I proposed
to wall in or wall out, because walls have two sides.
"Something there is that doesn't love a wall," wrote the
poet Robert Frost and I do believe he touched our national nerve. We are not a white country, we are a multi-
colored country. I am a Christian but this is not a Christian
country; it is a secular country of people of many faiths
and sometimes no faith. Our strength and power are born
of our diversity and tolerance. If we ever lose that, we will
have betrayed our legacy and our future will be compromised.

In New York City you will find communities and
neighborhoods of people from all over the world – Italian,
German, Indian, Jamaican, Norway, Somalia, Ghana, Puer-
to Rican, Chinese, Japanese, people of all colors and faiths
and cultural traditions. The city is a rich tapestry of America. When I am there, I inhale the busy city air and revel in

its energy. Other cities offer varieties of the same mix. It is who we are, what we are all about.

It is true that in many cities white people seal themselves off in enclaves apart from the racial and cultural cacophony around them, sending their children to private, all-white schools. They are making a grave mistake and doing a disservice to our country and their children. It is the mixture, and interaction of people from different backgrounds, that opens our minds and lifts us above the blind cultural prejudices that inhibit human progress in more backward nations.

By its very nature, democracy is chaotic. The first democracy in ancient Greece was crude – the citizens voted directly on just about everything of substance -- and hence subject to excesses that eventually brought it down. For thousands of years the educated people of the western world assumed democracy was unworkable, based primarily on the history of Greece. The architects of this republic – they did not call it a democracy – were careful to design a representative government that inhibited popular movements. The potential Achilles heel of democracy has always been the possibility of mob rule – that extreme situations would bring to power extreme leaders who would carelessly tear asunder our wonderful mosaic of people and cultures, catering to our prejudices and fears. We have sustained our heritage for almost two and a half centuries, but it hasn't always been easy and it is never a sure thing. During the Great Depression there were demagogues tearing at the fabric of our society, advocating extreme policies. Each new generation must relearn the basic lessons of history and embrace our shared heritage of diversity and tolerance.

That I believe is the ultimate challenge of the new digital technology. We must harness its power in positive ways, but be ever on guard to make certain it unites us instead of dividing us. We have to get into those ear sets the young people wear and teach them about our ancestors, what they stood for, what they bequeathed to us and how dearly it was purchased. When they hear demagogues preaching division, they need to know that siren song is small, stupid and unworthy of us. They need to know about the brave patriots from all backgrounds who have sacrificed their lives so that we might live in freedom. On a more fundamental level, we must teach our young people -- the next generation -- what real freedom is and that it does not occur naturally in this world. It depends on individuals endowed with character and patriotism who will stand against the mob in defense of the despised and downtrodden. That is what we are all about it. If it isn't, the American dream is a fraud and our decline will be of scant significance.

We need to get back to basics, working with governments on the state and local level, depending less on this Disneyland on the Potomac I live in. We need fewer prisons and more strong, intact families in which adults teach young people solid values and nourish them through their formative years. Government programs and social attitudes that reward and encourage illegitimacy are a real and present threat to our national character.

We need to wean the young people away from those electronic devices – at least some of the time. We should make it possible for everyone – rich and poor – to get out of the city, away from work and beyond the reach of e-mail and texts, where they can savor the wonders of our natural heritage, learn the joys of quiet reflection and actually converse with other people directly about the things we all care deeply about. Yes, have them sitting around the campfire singing Kum Bay Yah. We need a national program to provide summer camp to everyone with strict adult supervision and guidance We need a national draft that requires every able bodied young person to perform two years of public service of some kind.

I've always loved that Joe South song – "All God's children get weary when they roam, don't it make you want to go home." I have come a long way in my life, and I marvel at the extraordinary breakthroughs of the modern age. But sometimes I think we have come too far, too fast. I have read that then the first railroad trains became available to ordinary working people, many of them were freaked out by the speed at which they travelled – which may have been 20 or 30 miles an hour with those early machines. Up until then, the fastest speed they had known was a horse drawn conveyance of some kind.

The first trains came as a shock to people, even as the new technologies are coming as a shock to us today. We all need to slow down, take a deep breath and ponder the timeless verities of our ancestors, re-walk if only in our minds that long crooked road that got us where we are today. On some inner level, we all need to go home now and then.

## About The Author

LTG (Ret) Clarence E. "Mac" McKnight, Jr., a native of Tennessee and 1952 graduate of West Point, served in the Korean War, commanded two signal battalions in Europe, including commander, 5th Signal Command/deputy chief of staff for communications-electronics, U.S. Army Europe. He served as commandant of the U.S. Army Signal Center and School, and commander, U.S. Army Communications Command. Mac served tours as commander of Fort Gordon, Georgia, and Fort Huachuca, Arizona, and concluded his active career as Director of Command, Control and Communications Systems for the Joint Chiefs of Staff in Washington, D.C. McKnight is the author of "From Pigeons to Tweets: A General Who Led Dramatic Changes in Military Communications," published by History Publishing Company in 2013.

# ABOUT HANK H. COX

Hank H. Cox is a veteran reporter, freelance writer, speech writer and media relations manager. His articles and stories have appeared in a variety of publications. He is the author of "Lincoln and the Sioux Uprising of 1862," published by Cumberland House in 2005, and his upcoming book "The General Who Wore Six Stars: The Inside Story of Lt. Gen. John "Jesus Christ Himself" Lee," will be published by Potomac Books in fall 2017.